Student's Solutions Manual

To accompany

Contemporary Abstract Algebra
Sixth Edition

Student's Solutions Manual

To accompany

Contemporary Abstract Algebra
Sixth Edition

Joseph A. Gallian
University of Minnesota Duluth

Houghton Mifflin Company   Boston   New York

Publisher: Jack Shira
Sponsoring Editor: Lauren Schultz
Associate Editor: Lisa Collette
Editorial Associate: Kasey McGarrigle
Editorial Assistant: Brett Pasinella
Senior Manufacturing Coordinator: Chuck Dutton
Senior Marketing Manager: Ben Rivera
Marketing Associate: Lisa Lawler

Printed in the U.S.A.

ISBN: 0-618-54785-1

123456789-FFG-08 07 06 05 04

CONTENTS

Integers and Equivalence Relations

Groups

Fields

Special Topics

Student's Solutions Manual

To accompany

Contemporary Abstract Algebra
Sixth Edition

CHAPTER 0

Preliminaries

1. $\{1,2,3,4\}$; $\{1,3,5,7\}$; $\{1,5,7,11\}$; $\{1,3,7,9,11,13,17,19\}$;
$\{1,2,3,4,6,7,8,9,11,12,13,14,16,17,18,19,21,22,23,24\}$

3. 12, 2, 2, 10, 1, 0, 4, 5.

5. 1942, June 18; 1953, December 13

7. By using 0 as an exponent if necessary, we may write
$a = p_1^{m_1} \cdots p_k^{m_k}$ and $b = p_1^{n_1} \cdots p_k^{n_k}$, where the p's are distinct primes
and the m's and n's are nonnegative. Then $\mathrm{lcm}(a,b) = p_1^{s_1} \cdots p_k^{s_k}$,
where $s_i = \max(m_i, n_i)$ and $\gcd(a,b) = p_1^{t_1} \cdots p_k^{t_k}$, where
$t_i = \min(m_i, n_i)$ Then $\mathrm{lcm}(a,b) \cdot \gcd(a,b) = p_1^{m_1+n_1} \cdots p_k^{m_k+n_k} = ab$.

9. Write $a = nq_1 + r_1$ and $b = nq_2 + r_2$, where $0 \le r_1, r_2 < n$. We may
assume that $r_1 \ge r_2$. Then $a - b = n(q_1 - q_2) + (r_1 - r_2)$, where
$r_1 - r_2 \ge 0$. If $a \bmod n = b \bmod n$, then $r_1 = r_2$ and n divides $a - b$.
If n divides $a - b$, then by the uniqueness of the remainder, we then
have $r_1 - r_2 = 0$. Thus, $r_1 = r_2$ and therefore $a \bmod n = b \bmod n$.

11. By Exercise 9, to prove that $(a + b) \bmod n = (a' + b') \bmod n$ and
$(ab) \bmod n = (a'b') \bmod n$ it suffices to show that n divides
$(a + b) - (a' + b')$ and $ab - a'b'$. Since n divides both $a - a'$ and n
divides $b - b'$, it divides their difference. Because $a = a' \bmod n$ and
$b = b' \bmod n$ there are integers s and t such that $a = a' + ns$ and
$b = b' + nt$. Thus $ab = (a' + ns)(b' + nt) = a'b' + nsb' + a'nt + nsnt$.
Thus, $ab - a'b'$ is divisible by n.

13. Suppose that there is an integer n such that $ab \bmod n = 1$. Then
there is an integer q such that $ab - nq = 1$. Since d divides both a
and n, d also divides 1. So, $d = 1$. On the other hand, if $d = 1$,
then by Theorem 0.2, there are integers s and t such that
$as + nt = 1$. Thus, modulo n, $as = 1$.

15. Since st divides $a - b$, both s and t divide $a - b$. The converse is
true when $\gcd(s, t) = 1$.

17. If $\gcd(a, bc) = 1$, then there is no prime that divides both a and bc. By Euclid's Lemma and unique factorization, this means that there is no prime that divides both a and b or both a and c. Conversely, if no prime divides both a and b or both a and c, then by Euclid's Lemma, no prime divides both a and bc.

19. Suppose that there are only a finite number of primes $p_1, p_2, \ldots, p_n$. Then, by Exercise 18, $p_1 p_2 \ldots p_n + 1$ is not divisible by any prime. This means that $p_1 p_2 \ldots p_n + 1$, which is larger than any of $p_1, p_2, \ldots, p_n$, is itself prime. This contradicts the assumption that $p_1, p_2, \ldots, p_n$ is the list of all primes.

21. Let S be a set with $n + 1$ elements and pick some a in S. By induction, S has 2^n subsets that do not contain a. But there is one-to-one correspondence between the subsets of S that do not contain a and those that do. So there are $2 \cdot 2^n = 2^{n+1}$ subsets in all.

23. Consider $n = 200! + 2$. Then 2 divides n, 3 divides $n + 1$, 4 divides $n + 2, \ldots$, and 202 divides $n + 200$.

25. Say $p_1 p_2 \cdots p_r = q_1 q_2 \cdots q_s$, where the p's and the q's are primes. By the Generalized Euclid's Lemma, p_1 divides some q_i, say q_1 (we may relabel the q's if necessary). Then $p_1 = q_1$ and $p_2 \cdots p_r = q_2 \cdots q_s$. Repeating this argument at each step we obtain $p_2 = q_2, \cdots, p_r = q_r$ and $r = s$.

27. Suppose that S is a set that contains a and whenever $n \geq a$ belongs to S, then $n + 1 \in S$. We must prove that S contains all integers greater than or equal to a. Let T be the set of all integers greater than a that are not in S and suppose that T is not empty. Let b be the smallest integer in T (if T has no negative integers, b exists because of the Well Ordering Principle; if T has negative integers, it can have only a finite number of them so that there is a smallest one). Then $b - 1 \in S$, and therefore $b = (b - 1) + 1 \in S$. This contradicts our assumption that b is not in S.

29. The statement is true for any divisor of $8^4 - 4 = 4092$.

31. 6 p.m.

33. Observe that the number with the decimal representation $a_9 a_8 \ldots a_1 a_0$ is $a_9 \cdot 10^9 + a_8 \cdot 10^8 + \cdots + a_1 \cdot 10 + a_0$. From

Exercise 11 and the fact that $a_i 10^i \bmod 9 = a_i \bmod 9$ we deduce that the check digit is $(a_9 + a_8 + \cdots + a_1 + a_0) \bmod 9$. So, substituting 0 for 9 or vice versa for any a_i does not change the value of $(a_9 + a_8 + \cdots + a_1 + a_0) \bmod 9$.

35. For the case in which the check digit is not involved, the argument given Exercise 33 applies to transposition errors. Denote the money order number by $a_9 a_8 \ldots a_1 a_0 c$ where c is the check digit. For a transposition involving the check digit $c = (a_9 + a_8 + \cdots + a_0) \bmod 9$ to go undetected, we must have $a_0 = (a_9 + a_8 + \cdots + a_1 + c) \bmod 9$. Substituting for c yields $2(a_9 + a_8 + \cdots + a_0) \bmod 9 = a_0$. Then cancelling the a_0, multiplying by sides by 5, and reducing module 9, we have $10(a_9 + a_8 + \cdots + a_1) = a_9 + a_8 + \cdots + a_1 = 0$. It follows that $c = a_9 + a_8 \cdots + a_1 + a_0 = a_0$. In this case the transposition does not yield an error.

37. Say the number is $a_8 a_7 \ldots a_1 a_0 = a_8 10^8 + a_7 10^7 + \cdots + a_1 10 + a_0$. Then the error is undetected if and only if $(a_i 10^i - a_i' 10^i) \bmod 7 = 0$. Multiplying both sides by 5^i and noting that $50 \bmod 7 = 1$, we obtain $(a_i - a_i') \bmod 7 = 0$.

39. 4

43. Cases where $(2a - b - c) \bmod 11 = 0$ are undetected.

45. Since $10a_1 + 9a_2 + \cdots + a_{10} = 0 \bmod 11$ if and only if
$0 = (-10a_1 - 9a_2 - \cdots - 10a_{10}) \bmod 11 =$
$(a_1 + 2a_2 + \cdots + 10a_{10}) \bmod 11$, the check digit would be the same.

47. First note that the sum of the digits modulo 11 is 2. So, some digit is 2 too large. Say the error is in position i. Then
$10 = (4,3,0,2,5,1,1,5,6,8) \cdot (1,2,3,4,5,6,7,8,9,10) \bmod 11 = 2i$.
Thus, the digit in position 5 to 2 too large. So, the correct number is 4302311568.

49. No. $(1,0) \in R$ and $(0,-1) \in R$ but $(1,-1) \notin R$.

51. a belongs to the same subset as a. If a and b belong to the subset A and b and c belong to the subset B, then $A = B$, since the distinct subsets of P are disjoint. So, a and c belong to A.

53. 　2. Since β is one-to-one, $\beta(\alpha(a_1)) = \beta(\alpha(a_2))$ implies that $\alpha(a_1) = \alpha(a_2)$ and since α is one-to-one, $a_1 = a_2$.

3. Let $c \in C$. There is a b in B such that $\beta(b) = c$ and an a in A such that $\alpha(a) = b$. Thus, $(\beta\alpha)(a) = \beta(\alpha(a)) = \beta(b) = c$.

4. Since α is one-to-one and onto we may define $\alpha^{-1}(x) = y$ if and only if $\alpha(y) = x$. Then $\alpha^{-1}(\alpha(a)) = a$ and $\alpha(\alpha^{-1}(b)) = b$.

CHAPTER 1
Introduction to Groups

1. Three rotations: $0°$, $120°$, $240°$, and three reflections across lines from vertices to midpoints of opposite sides.

3. no

5. D_n has n rotations of the form $k(360°/n)$, where $k = 0, \ldots, n-1$. In addition, D_n has n reflections. When n is odd, the axes of reflection are the lines from the vertices to the midpoints of the opposite sides. When n is even, half of the axes of reflection are obtained by joining opposite vertices; the other half, by joining midpoints of opposite sides.

7. A rotation followed by a rotation either fixes every point (and so is the identity) or fixes only the center of rotation. However, a reflection fixes a line.

9. Observe that $1 \cdot 1 = 1$; $1(-1) = -1$; $(-1)1 = -1$; $(-1)(-1) = 1$. These relationships also hold when 1 is replaced by a "rotation" and -1 is replaced by a "reflection."

11. $HD = DV$ but $H \neq V$.

13. R_0, R_{180}, H, V

15. R_0, R_{180}, H, V

17. In each case the group is D_6.

19. cyclic

21. D_{28}

23. It would wobble violently.

CHAPTER 2
Groups

1. The set does not contain the identity; closure fails.

3. Under multiplication modulo 4, 2 does not have an inverse. Under multiplication modulo 5, $\{1, 2, 3, 4\}$ is closed, 1 is the identity, 1 and 4 are their own inverses, and 2 and 3 are inverses of each other. Modulo multiplication is associative.

5. $\begin{bmatrix} 9 & 9 \\ 10 & 8 \end{bmatrix}$

7. (a) $2a + 3b$; (b) $-2a + 2(-b + c)$; (c) $-3(a + 2b) + 2c = 0$

9. First note that if 1 is in H then by closure all five elements are in H. Then, since $\gcd(p^q, q^p) = 1$ and $\gcd(p + q, pq) = 1$, Theorem 0.2 shows that 1 belongs to H in all cases except case e.

11. The set is closed because $\det(AB) = (\det A)(\det B)$. Matrix multiplication is associative. $\begin{bmatrix} 1 & 0 \\ 0 & 1 \end{bmatrix}$ is the identity.

    $\begin{bmatrix} a & b \\ c & d \end{bmatrix}^{-1} = \begin{bmatrix} d & -b \\ -c & a \end{bmatrix}$. (Note that the inverse satisfies the condition required.)

13. Using closure and trail and error, we discover that $9 \cdot 74 = 29$ and 29 is not on the list.

15. For $n \geq 0$, we use induction. The case that $n = 0$ is trivial. Then note that $(ab)^{n+1} = (ab)^n ab = a^n b^n ab = a^{n+1} b^{n+1}$. For $n < 0$, note that $e = (ab)^0 = (ab)^n (ab)^{-n} = (ab)^n a^{-n} b^{-n}$ so that $a^n b^n = (ab)^n$. In a non-Abelian group $(ab)^n$ need not equal $a^n b^n$.

17. Suppose that G is Abelian. Then by Exercise 16, $(ab)^{-1} = b^{-1}a^{-1} = a^{-1}b^{-1}$. If $(ab)^{-1} = a^{-1}b^{-1}$ then by Exercise 16 $b^{-1}a^{-1} = (ab)^{-1} = a^{-1}b^{-1}$. Taking the inverse on both sides and using Exercise 16 yields $ab = ba$.

19. The case where $n = 0$ is trivial. For $n > 0$, note that
    $(a^{-1}ba)^n = (a^{-1}ba)(a^{-1}ba) \cdots (a^{-1}ba)$ (n terms). So, cancelling the
    consecutive a and a^{-1} terms gives $a^{-1}b^n a$. For $n < 0$, note that
    $e = (a^{-1}ba)^n(a^{-1}ba)^{-n} = (a^{-1}ba)^n(a^{-1}b^{-n}a)$ and solve for
    $(a^{-1}ba)^n$.

21. By closure we have $\{1, 3, 5, 9, 13, 15, 19, 23, 25, 27, 39, 45\}$.

23. Suppose x appears in a row labeled with a twice. Say $x = ab$ and
    $x = ac$. Then cancellation gives $b = c$. But we use distinct elements
    to label the columns.

25. Proceed as follows. By definition of the identity, we may complete
    the first row and column. Then complete row 3 and column 5 by
    using Exercise 23. In row 2 only c and d remain to be used. We
    cannot use d in position 3 in row 2 because there would then be
    two d's in column 3. This observation allows us to complete row 2.
    Then rows 3 and 4 may be completed by inserting the unused two
    elements. Finally, we complete the bottom row by inserting the
    unused column elements.

27. $axb = c$ implies that $x = a^{-1}(axb)b^{-1} = a^{-1}cb^{-1}$; $a^{-1}xa = c$ implies
    that $x = a(a^{-1}xa)a^{-1} = aca^{-1}$.

29. Since e is one solution it suffices to show that nonidentity solutions
    come in disjoints pairs. To this end note that if $x^3 = e$ and $x \neq e$,
    then $(x^{-1})^3 = e$ and $x \neq x^{-1}$. So if we can find one nonidentity
    solution we can find a second one. Now suppose that a and a^{-1} are
    nonidentity elements that satisfy $x^3 = e$ and b is a nonidentity
    element such that $b \neq a$ and $b \neq a^{-1}$ and $b^3 = e$. Then, as before,
    $(b^{-1})^3 = e$ and $b \neq b^{-1}$. Moreover, $b^{-1} \neq a$ and $b^{-1} \neq a^{-1}$. Thus,
    finding a third nonidentity solution gives a fourth one. Continuing
    in this fashion we see that we always have an even number of
    nonidentity solutions to the equation $x^3 = e$.

    To prove the second statement note that if $x^2 \neq e$, then $x^{-1} \neq x$
    and $(x^{-1})^2 \neq e$. So, arguing as in the preceding case we see that
    solutions to $x^2 \neq e$ come in disjoint pairs.

31. Observe that $a(a^{-1}b^{-1})b = e(a^{-1}b^{-1})(ba)$. Cancelling the term
    $(a^{-1}b^{-1})$ on both sides we obtain $ab = eba = ba$.

33. Since $a^2 = b^2 = (ab)^2 = e$, we have $aabb = abab$. Now cancel on left
    and right.

35. If n is not prime, we can write $n = ab$, where $1 < a < n$ and $1 < b < n$. Then a and b belong to the set $\{1, 2, \ldots, n\}$ but $0 = ab \bmod n$ does not.

37. Since

$$\begin{bmatrix} a & a \\ a & a \end{bmatrix} \begin{bmatrix} b & b \\ b & b \end{bmatrix} = \begin{bmatrix} 2ab & 2ab \\ 2ab & 2ab \end{bmatrix}.$$

and $2ab \neq 0$ we have closure; matrix multiplication is associative; from the product above we observe that the identity is $\begin{bmatrix} 1/2 & 1/2 \\ 1/2 & 1/2 \end{bmatrix}$ and that the inverse of $\begin{bmatrix} a & a \\ a & a \end{bmatrix}$ is $\begin{bmatrix} 1/(4a) & 1/(4a) \\ 1/(4a) & 1/(4a) \end{bmatrix}$.

CHAPTER 3
Finite Groups; Subgroups

1. $|Z_{12}| = 12; |U(10)| = 4; |U(12)| = 4; |U(20)| = 8; |D_4| = 8.$
   In $Z_{12}, |0| = 1; |1| = |5| = |7| = |11| = 12; |2| = |10| = 6; |3| = |9| = 4; |4| = |8| = 3; |6| = 2.$
   In $U(10), |1| = 1; |3| = |7| = 4; |9| = 2.$
   In $U(20), |1| = 1; |3| = |7| = |13| = |17| = 4; |9| = |11| = |19| = 2.$
   In $D_4, |R_0| = 1; |R_{90}| = |R_{270}| = 4;$
   $|R_{180}| = |H| = |V| = |D| = |D'| = 2.$
   In each case, notice that the order of the element divides the order of the group.

3. In $Q, |0| = 1.$ All other elements have infinite order since $x + x + \cdots + x = 0$ only when $x = 0.$

5. In $Z_{30}, 2 + 28 = 0$ and $8 + 22 = 0.$ So, 2 and 28 are inverses of each other and 8 and 22 are inverses of each other. In $U(15), 2 \cdot 8 = 1$ and $7 \cdot 13 = 1.$ So, 2 and 8 are inverses of each other and 7 and 13 are inverses of each other.

7. If a has infinite order, then $e, a, a^2, \ldots$ are all distinct and belong to G, so G is infinite. If $|a| = n$, then $e, a, a^2, \ldots, a^{n-1}$ are all distinct and belong to G, so G has at least n elements.

9. Since $|U(20)| = 8$, for $U(20) = \langle k \rangle$ for some k it must the case that $|k| = 8.$ But $1^1 = 1, 3^4 = 1, 7^4 = 1, 9^2 = 1, 11^2 = 1, 13^4 = 1,$ $17^4 = 1,$ and $19^2 = 1.$ So, the maximum order of any element is 4.

11. In D_3, let R_{120} be a rotation of 120 degrees and F be any reflection. Then $R_{120}F$ is a reflection. Also, $(R_{120}F)F = R_{120}(FF) = R_{120}$ so that $|R_{120}F| = 2$ and $|F| = 2$ while $|(R_{120}F)F| = 3.$ In D_4, let R_{90} be a rotation of 90 degrees and F be any reflection. Then $R_{90}F$ is a reflection. Also, $(R_{90}F)F = R_{90}(FF) = R_{90}$ so that $|R_{90}F| = 2$ and $|F| = 2$ while $|(R_{90}F)F| = 4.$ In D_5, let R_{72} be a rotation of 72 degrees and F be any reflection. Then $R_{72}F$ is a reflection. Also,

$(R_{72}F)F = R_{72}(FF) = R_{72}$ so that $|R_{72}F| = 2$ and $|F| = 2$ while $|(R_{72}F)F| = 5$.

13. $U_4(20) = \{1, 9, 13, 17\}$; $U_5(20) = \{1, 11\}$; $U_5(30) = \{1, 11\}$; $U_{10}(30) = \{1, 11\}$. $U_k(n)$ is closed because (ab) mod $k = (a$ mod $k)(b$ mod $k) = 1 \cdot 1 = 1$ (here we used the fact that k divides n). H is not closed since $7 \in H$ but $7 \cdot 7 = 9$ is not in H.

15. If $x \in Z(G)$, then $x \in C(a)$ for all a, so $x \in \bigcap_{a \in G} C(a)$. If $x \in \bigcap_{a \in G} C(a)$, then $xa = ax$ for all a in G, so $x \in Z(G)$.

17.   **a.** $C(5) = G$; $C(7) = \{1, 3, 5, 7\}$

      b. $Z(G) = \{1, 5\}$

      c. $|2| = 2$; $|3| = 4$. They divide the order of the group.

19. Since $ea = ae, C(a) \neq \emptyset$. Suppose that x and y are in $C(a)$. Then $xa = ax$ and $ya = ay$. Thus,

$$(xy)a = x(ya) = x(ay) = (xa)y = (ax)y = a(xy)$$

and therefore $xy \in C(a)$. Starting with $xa = ax$, we multiply both sides by x^{-1} on the right and left to obtain $x^{-1}xax^{-1} = x^{-1}axx^{-1}$ and so $ax^{-1} = x^{-1}a$. This proves that $x^{-1} \in C(a)$. By the Two-Step Subgroup Test, $C(a)$ is a subgroup of G.

21. No. In D_4, $C(R_{180}) = D_4$.

23. Let $H = \{x \in G| \; x^n = e\}$. Since $e^1 = e, H \neq \emptyset$. Now let $a, b \in H$. Then $a^n = e$ and $b^n = e$. So, $(ab)^n = a^n b^n = ee = e$ and therefore $ab \in H$. Starting with $a^n = e$ and taking the inverse of both sides, we get $(a^n)^{-1} = e^{-1}$. This simplifies to $(a^{-1})^n = e$. Thus, $a^{-1} \in H$. By the Two-Step test, H is a subgroup of G. In D_4, $\{x| \; x^2 = e\} = \{R_0, R_{180}, H, V, D, D'\}$. This set is not closed because $HD = R_{90}$.

25. For any positive integer n the only solutions to $x^n = 1$ are $x = 1$ and $x = -1$. So, the only finite subgroups of $\mathbf{R}^*$ are $\{1\}$ and $\{1, -1\}$.

27. Note that $(ab)^2 = abab = a(ba)b = a(a^3b)b = a^4b^2 = ee = e$. So, $|ab| = 2$.

29. By induction we will prove that any positive integer n we have

$$\begin{bmatrix} 1 & 1 \\ 0 & 1 \end{bmatrix}^n = \begin{bmatrix} 1 & n \\ 0 & 1 \end{bmatrix}.$$

The $n = 1$ case is true by definition. Now assume

$$\begin{bmatrix} 1 & 1 \\ 0 & 1 \end{bmatrix}^k = \begin{bmatrix} 1 & k \\ 0 & 1 \end{bmatrix}.$$

Then

$$\begin{bmatrix} 1 & 1 \\ 0 & 1 \end{bmatrix}^{k+1} = \begin{bmatrix} 1 & 1 \\ 0 & 1 \end{bmatrix}^k \begin{bmatrix} 1 & 1 \\ 0 & 1 \end{bmatrix} =$$

$$\begin{bmatrix} 1 & k \\ 0 & 1 \end{bmatrix}\begin{bmatrix} 1 & 1 \\ 0 & 1 \end{bmatrix} = \begin{bmatrix} 1 & k+1 \\ 0 & 1 \end{bmatrix}.$$

So, when the entries are from $\mathbf{R}$, $\begin{bmatrix} 1 & 1 \\ 0 & 1 \end{bmatrix}$ has finite order. When the entries are from Z_p, the order is p.

31. For any positive integer n, a rotation of $360°/n$ has order n. If we let R be a rotation of $\sqrt{2}$ degrees then R^n is a rotation of $\sqrt{2}n$ degrees. This is never a multiple of $360°$, for if $\sqrt{2}n = 360k$ then $\sqrt{2} = 360k/n$, which is rational. So, R has infinite order.

33. $\langle R_0 \rangle, \langle R_{180} \rangle, \langle R_{90} \rangle, \langle D \rangle, \langle D' \rangle, \langle H \rangle, \langle V \rangle$. (Note that $\langle R_{90} \rangle = \langle R_{270} \rangle$.) The subgroups $\{R_0, R_{180}, D, D'\}$ and $\{R_0, R_{180}, H, V\}$ are not cyclic.

35. Nonidentity elements of odd order come in pairs. So, there must be some element a of even order, say $|a| = 2m$. Then $|a^m| = 2$.

37. Let $|g| = m$ and write $m = nq + r$ where $0 \le r < n$. Then $g^r = g^{m-nq} = g^m(g^n)^{-q}$ belongs to H. So, $r = 0$.

39. $1 \in H$, so $H \neq \emptyset$. Let $a, b \in H$. The $(ab^{-1})^2 = a^2(b^2)^{-1}$, which is the product of two rationals. The integer 2 can be replaced by any positive integer.

41. $|\langle 3 \rangle| = 4$

43. Let $\begin{bmatrix} a & b \\ c & d \end{bmatrix}$ and $\begin{bmatrix} a' & b' \\ c' & d' \end{bmatrix}$ belong to H. By the One-Step

Subgroup Test it suffices to show that
$a - a' + b - b' + c - c' + d - d' = 0$. This follows from
$a + b + c + d = 0 = a' + b' + c' + d'$. If 0 is replaced by 1, H is not a
subgroup since it does not contain the identity.

45. If 2^a and $2^b \in K$, then $2^a(2^b)^{-1} = 2^{a-b} \in K$, since $a - b \in H$.

47. $\begin{bmatrix} 2 & 0 \\ 0 & 2 \end{bmatrix}^{-1} = \begin{bmatrix} \frac{1}{2} & 0 \\ 0 & \frac{1}{2} \end{bmatrix}$ is not in H.

49. If $a + bi$ and $c + di \in H$, then
$(a+bi)(c+di)^{-1} = \frac{a+bi}{c+di}\frac{c-di}{c-di} = \frac{(ac+bd)+(bc-ad)i}{c^2+d^2} = (ac+bd)+(bc-ad)i.$
Moreover,
$(ac + bd)^2 + (bc - ad)^2 = a^2c^2 + 2acbd + b^2d^2 + b^2c^2 - 2bcad + a^2d^2.$
Simplifying we obtain,
$(a^2 + b^2)c^2 + (a^2 + b^2)d^2 = (a^2 + b^2)(c^2 + d^2) = 1 \cdot 1 = 1.$ So, H is a
subgroup. H is the unit circle in the complex plane.

51. **a.** Suppose that $\begin{bmatrix} a & b \\ c & d \end{bmatrix}$ commutes with $\begin{bmatrix} 1 & 1 \\ 1 & 0 \end{bmatrix}$. Then

$\begin{bmatrix} a & b \\ c & d \end{bmatrix}\begin{bmatrix} 1 & 1 \\ 1 & 0 \end{bmatrix} = \begin{bmatrix} 1 & 1 \\ 1 & 0 \end{bmatrix}\begin{bmatrix} a & b \\ c & d \end{bmatrix}$. This implies that

$\begin{bmatrix} a+b & a \\ c+d & c \end{bmatrix} = \begin{bmatrix} a+c & b+d \\ a & b \end{bmatrix}$. From this we have $b = c$ and

$a = b + d$. For convenience, we solve the latter equality for d in

terms of a and b and get $d = a - b$. So, $C\left(\begin{bmatrix} 1 & 1 \\ 1 & 0 \end{bmatrix}\right) =$

$\left\{\begin{bmatrix} a & b \\ b & a-b \end{bmatrix} \mid \text{where } a^2 - ab - b^2 \neq 0; a, b \in \mathbf{R}\right\}.$

b. Suppose that $\begin{bmatrix} a & b \\ c & d \end{bmatrix}$ commutes with $\begin{bmatrix} 0 & 1 \\ 1 & 0 \end{bmatrix}$. Then

$\begin{bmatrix} a & b \\ c & d \end{bmatrix}\begin{bmatrix} 0 & 1 \\ 1 & 0 \end{bmatrix} = \begin{bmatrix} 0 & 1 \\ 1 & 0 \end{bmatrix}\begin{bmatrix} a & b \\ c & d \end{bmatrix}.$

This implies that

$\begin{bmatrix} b & a \\ d & c \end{bmatrix} = \begin{bmatrix} c & d \\ a & b \end{bmatrix}.$

From this we have $b = c$ and $a = d$. So,

$$C\left(\begin{bmatrix} 0 & 1 \\ 1 & 0 \end{bmatrix}\right) = \left\{\begin{bmatrix} a & b \\ b & a \end{bmatrix} \mid \text{where } a^2 - b^2 \neq 0; a, b \in \mathbf{R}\right\}.$$

c. Suppose that $\begin{bmatrix} a & b \\ c & d \end{bmatrix}$ commutes with every element of $GL(2, \mathbf{R})$. By part b we know that $c = b$ and $d = a$. So, any element in the center of $GL(2, \mathbf{R})$ has the form

$$\begin{bmatrix} a & b \\ b & a \end{bmatrix}.$$

Also,

$$\begin{bmatrix} a & b \\ b & a \end{bmatrix}\begin{bmatrix} 1 & 0 \\ 0 & 0 \end{bmatrix} = \begin{bmatrix} 1 & 0 \\ 0 & 0 \end{bmatrix}\begin{bmatrix} a & b \\ b & a \end{bmatrix}.$$

This gives

$$\begin{bmatrix} a & 0 \\ b & 0 \end{bmatrix} = \begin{bmatrix} a & b \\ 0 & 0 \end{bmatrix}.$$

So, $b = 0$. Thus,

$$Z(G) = \left\{\begin{bmatrix} a & 0 \\ 0 & a \end{bmatrix} \mid a \neq 0; a \in \mathbf{R}\right\}.$$

53. By Theorem 0.2 there are integers s and t so that $1 = ms + nt$. Then $a^1 = a^{ms+nt} = a^{ms}a^{nt} = (a^m)^s(a^n)^t = (a^t)^n$.

CHAPTER 4
Cyclic Groups

1. For Z_6, generators are 1 and 5; for Z_8 generators are 1, 3, 5, and 7; for Z_{20} generators are 1, 3, 7, 9, 11, 13, 17, and 19.

3. $\langle 20 \rangle = \{20, 10, 0\}$
   $\langle 10 \rangle = \{10, 20, 0\}$

5. $\langle 3 \rangle = \{3, 9, 7, 1\}$
   $\langle 7 \rangle = \{7, 9, 3, 1\}$

7. U_8 or D_3.

9. Six subgroups; generators are the divisors of 20.
   Six subgroups; generators are a^k, where k is a divisor of 20.

11. By definition, $a^{-1} \in \langle a \rangle$. So, $\langle a^{-1} \rangle \subseteq \langle a \rangle$. By definition, $a = (a^{-1})^{-1} \in \langle a^{-1} \rangle$. So, $\langle a \rangle \subseteq \langle a^{-1} \rangle$.

13. Observe that $\langle a^{21} \rangle = \{e, a^{21}, a^{18}, a^{15}, a^{12}, a^9, a^6, a^3\}$ and $\langle a^{10} \rangle = \{e, a^{10}, a^{20}, a^6, a^{16}, a^2, a^{12}, a^{22}, a^8, a^{18}, a^4, a^{14}\}$. Since the intersection of two subgroups is a subgroup, according to the proof of Theorem 4.3, we can find a generator of the intersection by taking the smallest positive power of a that is in the intersection. So, $\langle a^{21} \rangle \cap \langle a^{10} \rangle = \langle a^6 \rangle$. For the case $\langle a^m \rangle \cap \langle a^n \rangle$, let $k = \text{lcm}(m, n)$. Write $k = ms$ and $k = nt$. Then $a^k = (a^m)^s \in \langle a^m \rangle$ and $a^k = (a^n)^t \in \langle a^n \rangle$. So, $\langle a^k \rangle \subseteq \langle a^m \rangle \cap \langle a^n \rangle$. Now let a^r be any element in $\langle a^m \rangle \cap \langle a^n \rangle$. Then r is a multiple of both m and n. It follows that r is a multiple of k (see Exercise 12 of Chapter 0). So, $a^r \in \langle a^k \rangle$.

15. $|g|$ divides 12 is equivalent to $g^{12} = e$. So, if $a^{12} = e$ and $b^{12} = e$, then $(ab^{-1})^{12} = a^{12}(b^{12})^{-1} = ee^{-1} = e$. The same argument works when 12 is replaced by any integer (see Exercise 23 of Chapter 3).

17. is odd or infinite

19. $\langle 1 \rangle$, $\langle 7 \rangle$, $\langle 11 \rangle$, $\langle 17 \rangle$, $\langle 19 \rangle$, $\langle 29 \rangle$

21. **a.** $|a|$ divides 12.

    b. $|a|$ divides m.

    c. By Theorem 4.3, $|a| = 1, 2, 3, 4, 6, 8, 12$, or 24. If $|a| = 2$, then $a^8 = (a^2)^4 = e^4 = e$. A similar argument eliminates all other possibilities except 24.

23. Yes, by Theorem 4.3. The subgroups of Z are of the form $\langle n \rangle = \{0, \pm n, \pm 2n, \pm 3n, \ldots\}$, where n is any integer.

25. D_n has n reflections each of which has order 2. D_n also has n rotations that form a cyclic group of order n. So, according to Theorem 4.4, there are $\phi(d)$ rotations of order d in D_n. If n is odd, there are no rotations of order 2. If n is even, there is $\phi(2) = 1$ rotation of order 2. (Namely, R_{180}.) So, when n is odd D_n has n elements of order 2; when n is even, D_n has $n + 1$ elements of order 2.

27. Suppose that a^k generates $\langle a \rangle$. Then there is an integer t so that $(a^k)^t = a$. By Theorem 4.1, we conclude that $kt = 1$. So, $k = \pm 1$.

29. 1000000, 3000000, 5000000, 7000000. By Theorem 4.3, $\langle 1000000 \rangle$ is the unique subgroup of order 8, and only those on the list are generators.

31. Let $G = \{a_1, a_2, \ldots, a_k\}$. Now let $|a_i| = n_i$ and $n = n_1 n_2 \ldots n_k$. Then $a_i^n = e$ for all i since n is a multiple of n_i.

33.

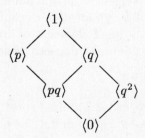

35.

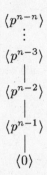

$$\langle p^{n-n} \rangle$$
$$\vdots$$
$$\langle p^{n-3} \rangle$$
$$|$$
$$\langle p^{n-2} \rangle$$
$$|$$
$$\langle p^{n-1} \rangle$$
$$|$$
$$\langle 0 \rangle$$

37. Suppose a and b are relatively prime positive integers and $\langle a/b \rangle = Q^+$. Then there is some positive integer n such that $(a/b)^n = 2$. Clearly $n \neq 0, 1,$ or -1. If $n > 1$, $a^n = 2b^n$, so that 2 divides a. But then 2 divides b as well. A similar contradiction occurs if $n < -1$.

39. For 6, use Z_{25}. For n, use $Z_{2^{n-1}}$.

41. Let $t = \text{lcm}(m, n)$ and let $|ab| = s$. Then $(ab)^t = a^t b^t = e$, and therefore s divides t. Also, $e = (ab)^s = a^s b^s$, so that $a^s = b^{-s}$, and therefore a^s and b^{-s} belong to $\langle a \rangle \cap \langle b \rangle = \{e\}$. Thus, m divides s and n divides s, and therefore t divides s.

For the second part, let R_{120} denote a rotation of $120°$ and F be a reflection in D_3. Then $\langle R_{120} \rangle \cap \langle F \rangle = \{R_0\}$, $|R_{120}| = 3$, $|F| = 2$ but D_3 has no element of order 6.

43. An infinite cyclic group does not have element of prime order. A finite cyclic group can have only one subgroup for each divisor of its order. A subgroup of order p has exactly $p - 1$ elements of order p. Another element of order p would give another subgroup of order p.

45. $1 \cdot 4, 3 \cdot 4, 7 \cdot 4, 9 \cdot 4$

47. D_{33} has 33 reflections each of which has order 2 and 33 rotations that form a cyclic group. So, according to Theorem 4.4, for each divisor d of 33 there are $\phi(d)$ rotations of order d in D_n. This gives one element of order 1; $\phi(3) = 2$ elements of order 3; $\phi(11) = 10$ elements of order 11; and $\phi(33) = 20$ elements of order 33.

49. Let $|\langle a \rangle| = 4$ and $|\langle b \rangle| = 5$. Since $(ab)^{20} = (a^4)^5(b^5)^4 = e \cdot e = e$ we know that $|ab|$ divides 20. Noting that $(ab)^4 = b^4 \neq e$ we know that $|ab| \neq 1, 2$ or 4. Likewise, $(ab)^{10} = a^2 \neq e$ implies that $|ab| \neq 5$ or 10. So, $|ab| = 20$. Then, by Theorem 4.3, $\langle ab \rangle$ has subgroups of orders 1, 2, 4, 5, 10 and 20.

51. Say a and b are distinct elements of order 2. If a and b commute, then ab is a third element of order 2. If a and b do not commute, then aba is a third element of order 2.

53. By Exercise 14 of Chapter 3, $\langle a \rangle \cap \langle b \rangle$ is a subgroup. Also, $\langle a \rangle \cap \langle b \rangle \subseteq \langle a \rangle$ and $\langle a \rangle \cap \langle b \rangle \subseteq \langle b \rangle$. So, by Theorem 4.3, $|\langle a \rangle \cap \langle b \rangle|$ is a common divisor of 10 and 21. Thus, $|\langle a \rangle \cap \langle b \rangle| = 1$ and therefore $\langle a \rangle \cap \langle b \rangle = \{e\}$.

55. Similar to Exercise 53, $|\langle a \rangle \cap \langle b \rangle|$ must divide both 24 and 10. So, $|\langle a \rangle \cap \langle b \rangle| = 1$ or 2.

57. If $x^8 = e$, then $|x|$ divides 8. If G is cyclic, then by Theorem 4.4, G has exactly $\phi(8) + \phi(4) + \phi(2) + \phi(1) = 4 + 2 + 1 + 1 = 8$ elements with orders that divide 8. So, G is not cyclic. For the case $x^4 = e$, if G is cyclic, then G has $\phi(4) + \phi(2) + \phi(1) = 4$ elements with orders that divide 4. So, again G is not cyclic. If G is a finite cyclic group and n divides $|G|$, then there are exactly n elements in G that are solutions of $x^n = e$. So, if there are more than n elements x such that $x^n = e$, the group is not cyclic.

59. Observe that $|a^5| = 12$ implies that $e = (a^5)^{12} = a^{60}$ so $|a|$ divides 60. Since $\langle a^5 \rangle \subseteq \langle a \rangle$ we know that $|\langle a \rangle|$ is divisible by 12. So, $|\langle a \rangle| = 12$ or 60.

If $|a^4| = 12$, then $|a|$ divides 48. Since $\langle a^4 \rangle \subseteq \langle a \rangle$ we know that $|\langle a \rangle|$ is divisible by 12. So, $|\langle a \rangle| = 12, 24$, or 48. But $|a| = 12$ implies $|a^4| = 3$ and $|a| = 24$ implies $|a^4| = 6$. So, $|a| = 48$.

61. Say b is a generator of the group. Since p and $p^n - 1$ are relatively prime, we know by Corollary 2 of Theorem 4.2 that b^p also generates the group. Finally, observe that $(b^p)^k = (b^k)^p$.

63. By Exercise 14 of Chapter 3, $\langle a \rangle \cap \langle b \rangle$ is a subgroup $\langle a \rangle$ and $\langle b \rangle$. So, $|\langle a \rangle \cap \langle b \rangle|$ divides 12 and 22. If follows that $|\langle a \rangle \cap \langle b \rangle| = 1$ or 2 and since $\langle a \rangle \cap \langle b \rangle \neq \{e\}$ we have that $|\langle a \rangle \cap \langle b \rangle| = 2$. Because $\langle a^6 \rangle$ is the only subgroup of $\langle a \rangle$ of order 2 and $\langle b^{11} \rangle$ is the only subgroup of $\langle b \rangle$ of order 2, we have $\langle a \rangle \cap \langle b \rangle = \langle a^6 \rangle = \langle b^{11} \rangle$ and therefore $a^6 = b^{11}$.

65. The set is closed because the coefficients form a closed set under addition modulo 3. Polynomial addition is associative because the addition is term wise and the set $\{0, 1, 2\}$ under modulo 3 addition is associative. The identity is the polynomial $0x^2 + 0x + 0$. The inverse of $ax^2 + bx + c$ is $a'x^2 + b'x + c'$ where a', b' and c' are the inverses of a, b, and c modulo 3. Since there are 3 choices for each of a, b, and c, the group has 27 elements. Noting that adding any polynomial in the group to itself 3 times results in a polynomial with coefficients 0 modulo 3, we see that the largest order of any element is 3 so G is not cyclic.

SUPPLEMENTARY EXERCISES FOR CHAPTERS 1-4

1.   **a.** Let xh_1x^{-1} and xh_2x^{-1} belong to xHx^{-1}. Then
     $(xh_1x^{-1})(xh_2x^{-1})^{-1} = xh_1h_2^{-1}x^{-1} \in xHx^{-1}$ also.

   b. Let $\langle h \rangle = H$. Then $\langle xhx^{-1} \rangle = xHx^{-1}$.

   c. $(xh_1x^{-1})(xh_2x^{-1}) = xh_1h_2x^{-1} = xh_2h_1x^{-1} = (xh_2x^{-1})(xh_1x^{-1})$.

3. Suppose $\mathrm{cl}(a) \cap \mathrm{cl}(b) \neq \emptyset$. Say $xax^{-1} = yby^{-1}$. Then
   $(y^{-1}x)a(y^{-1}x)^{-1} = b$. Thus, for any ubu^{-1} in $\mathrm{cl}(b)$, we have
   $ubu^{-1} = uy^{-1}xa(y^{-1}x)^{-1}u^{-1} = (uy^{-1}x)a(uy^{-1}x)^{-1} \in \mathrm{cl}(a)$. This
   shows that $\mathrm{cl}(b) \subseteq \mathrm{cl}(a)$. By symmetry, $\mathrm{cl}(a) \subseteq \mathrm{cl}(b)$. Because
   $a = eae^{-1} \in \mathrm{cl}(a)$, the union of the conjugacy classes is G.

5. Observe that $e = (xax^{-1})^k = xa^kx^{-1}$ if and only if $a^k = e$.

7. In D_4, take $a = R_{90}$, $b = H$, and $c = D$.

9. By Exercise 5, for every x in G $|xax^{-1}| = |a|$, so that $xax^{-1} = a$ or
   $xa = ax$.

11. 1 of order 1, 15 of order 2, 8 of order 15, 4 of order 5, 2 of order 3.

13. Let $|G| = 5$. Let $a \neq e$ belong to G. If $|a| = 5$, we are done. If
   $|a| = 3$, then $\{e, a, a^2\}$ is a subgroup of G. Let b be either of the
   remaining two elements of G. Then the set $\{e, a, a^2, b, ab, a^2b\}$
   consists of six different elements, a contradiction. Thus $|a| \neq 3$.
   Similarly $|a| \neq 4$. We may now assume that every nonidentity
   element of G has order 2. Pick $a \neq e$ and $b \neq e$ in G with $a \neq b$.
   Then $\{e, a, b, ab\}$ is a subgroup of G. Let c be the remaining
   element of G. Then $\{e, a, b, ab, c, ac, bc, abc\}$ is a set of eight distinct
   elements of G, a contradiction. It now follows that if $a \in G$ and
   $a \neq e$, then $|a| = 5$.

15. $a^n(b^n)^{-1} = (ab^{-1})^n$, so by the One-Step Subgroup Test G^n is a
   subgroup. For the non-Abelian group, use D_3 with $n = 3$.

17. Suppose $G = H \cup K$. Pick $h \in H$ with $h \notin K$. Pick $k \in K$ with $k \notin H$. Then, $hk \in G$ but $hk \notin H$ and $hk \notin K$.
$U(8) = \{1, 3\} \cup \{1, 5\} \cup \{1, 7\}$.

19. If $|a| = p^k$ and $|b| = p^r$ with $k \leq r$, say, then $(ab^{-1})^{p^r} = a^{p^r}(b^{p^r})^{-1} = e$ so $|ab^{-1}|$ divides p^r.

21.

	e	a	a^2	b	ab	a^2b
e	e	a	a^2	b	ab	a^2b
a	a	a^2	e	ab	a^2b	b
a^2	a^2	e	a	a^2b	b	ab
b	b	a^2b	ab	e	a^2	a
ab	ab	b	a^2b	a	e	a^2
a^2b	a^2b	ab	b	a^2	a	e

D_3 satisfies these conditions.

23. $xy = yx$ if and only if $xyx^{-1}y^{-1} = e$. But $(xy)x^{-1}y^{-1} = x^{-1}(xy)y^{-1} = ee = e$.

25. Let $x \in N(gHg^{-1})$. Then $x(gHg^{-1})x^{-1} = gHg^{-1}$. Thus $g^{-1}xgHg^{-1}x^{-1}g = g^{-1}xgH(g^{-1}xg)^{-1} = H$. This means that $g^{-1}xg \in N(H)$. So $x \in gN(H)g^{-1}$. To show $gN(H)g^{-1} \subseteq N(gHg^{-1})$ let $x = gng^{-1}$ where $n \in N(H)$. Then $x(gHg^{-1})x^{-1} = (gng^{-1})(gHg^{-1})(gn^{-1}g^{-1}) = gnHn^{-1}g^{-1} = gHg^{-1}$. So, $x \in N(gHg^{-1})$.

27. In D_{11} let a be a reflection and let b be a rotation other than the identity. Then $|a| = 2$, $|b| = 11$ and $|ab| = 2$ because ab is a reflection.

29. Solution from *Mathematics Magazine*.[1] "Yes. Let a be an arbitrary element of S. The set $\{a^n \mid n = 1, 2, 3, \ldots\}$ is finite, and therefore $a^m = a^n$ for some m, n with $m > n \geq 1$. By cancellation we have $a^{r(a)} = a$, where $r(a) = m - n + 1 > 1$. If x is any element of S, then $aa^{r(a)-1}x = a^{r(a)}x = ax$, and this implies that $a^{r(a)-1}x = x$. Similarly, we see that $xa^{r(a)-1} = x$, and therefore the element $e = a^{(r(a)-1}$ is an identity. The identity element is unique, for e' is another identity, then $e = ee' = e'$. If $r(a) > 2$ then $a^{r(a)-2}$ is an inverse of a, and if $r(a) = 2$ then $a^2 = a = e$ is its own inverse. Thus S is a subgroup."

[1] *Mathematics Magazine* 63 (April 1990): 136.

31. 1^1 is rational so $H \neq \emptyset$. Say a^m and b^n are rational. Then $(ab^{-1})^{mn} = (a^m)^n/(b^n)^m$ is rational.

33. H contains the identity. If $A, B \in H$, then $\det(AB^{-1}) = (\det A)(\det B)^{-1}$ is rational. H is not a subgroup when $\det A$ is an integer, since $\det A^{-1} = (\det A)^{-1}$ is an integer only when $\det A = \pm 1$.

35. First suppose that G is not cyclic. Choose $x \neq e$ and choose $y \notin \langle x \rangle$. Then, since G has only two proper subgroups $G = \langle x \rangle \cup \langle y \rangle$. But then $xy \in \langle y \rangle$ so that $\langle x \rangle \subseteq \langle y \rangle$ and therefore $G = \langle y \rangle$ and G is cyclic. To prove that $|G| = pq$, where p and q are distinct primes or $|G| = p^3$, where p is prime we first observe that G is not infinite since an infinite cyclic group has infinitely many subgroups. If $|G|$ is divisible by at least two distinct primes p and q, then $|G| = pqm$. If $m > 1$, then G would have nontrivial proper subgroups of order p, q, and m. So, $m = 1$. If $|G| = p^n$, where p is prime, then by Theorem 4.3, G has exactly $n - 1$ nontrivial proper subgroups and therefore $n = 3$.

37. If T and U are not closed, then there are elements x and y in T and w and z in U such that xy is not in T and wz is not in U. It follows that $xy \in U$ and $wz \in T$. Then $xywz = (xy)wz \in U$ and $xywz = xy(wz) \in T$, a contradiction since T and U are disjoint.

39. Let G be the group of all polynomials with integer coefficients under addition. Let H_k be the subgroup of polynomials of degree at most k together with the zero polynomial (the zero polynomial does not have a degree).

41. Take $g = a$.

43. Let $S = \{s_1, s_2, s_3, \ldots, s_k\}$ and let g be any element in G. Then the set $\{gs_1^{-1}, gs_2^{-1}, gs_3^{-1}, \ldots, gs_k^{-1}\}$ consists of k distinct elements so it and S have at least one element in common. Say $gs_i^{-1} = s_j$. Then $g = s_j s_i$.

45. Let $K = \{x \in G \mid |x| \text{ divides } d\}$. The subgroup test shows that K is a subgroup. Let $x \in H$. By Theorem 4.3 $|x|$ divides d. So, $H \subseteq K$. Let $y \in K$, $|y| = t$ and $d = tq$. By Theorem 4.3, H has a subgroup of order t and G has only one subgroup of order t. So, $\langle y \rangle \subseteq H$.

CHAPTER 5
Permutation Groups

1. a. 2; b. 3; c. 5

3.　a. By Theorem 5.3 the order is lcm(3,3) = 3.

　　b. By Theorem 5.3 the order is lcm(3,4) = 12.

　　c. By Theorem 5.3 the order is lcm(3,2) = 6.

　　d. By Theorem 5.3 the order is lcm(3,6) = 6.

　　e. $|(1235)(24567)| = |(124)(3567)| = $ lcm(3,4) = 12.

　　f. $|(345)(245)| = |(25)(34)| = $ lcm(2,2) = 2.

5. By Theorem 5.3 the order is lcm(4,6) = 12

7. We find the orders by looking at the possible products of disjoint cycles.

(6-cycle) has order 6 and is odd;

(5-cycle)(1-cycle) has order 5 and is even;

(4-cycle)(2-cycle) has order 4 and is even;

(4-cycle)(1-cycle)(1-cycle) has order 4 and is odd;

(3-cycle)(3-cycle) has order 3 and is even;

(3-cycle)(2-cycle)(1-cycle) has order 6 and is odd;

(3-cycle)(1-cycle)(1-cycle)(1-cycle) has order 3 and is even;

(2-cycle)(2-cycle)(2-cycle) has order 2 and is odd;

(2-cycle)(2-cycle)(1-cycle)(1-cycle) has order 2 and is even;

(2-cycle)(1-cycle)(1-cycle)(1-cycle)(1-cycle) has order 2 and is odd.

So, for S_6, the possible orders are 1, 2, 3, 4, 5, 6; for A_6 the possible orders are 1, 2, 3, 4, 5.

For S_7 the possible cycle structures are:

(7-cycle) has order 7 and is even;

(6-cycle)(1-cycle) has order 6 and is odd;

(5-cycle)(2-cycle) has order 10 and is odd;

(5-cycle)(1-cycle)(1-cycle) has order 5 and is even;

(4-cycle)(3-cycle) has order 12 and is odd;

(4-cycle)(2-cycle)(1-cycle) has order 4 and is even;
(4-cycle)(1-cycle)(1-cycle)(1-cycle) has order 4 and is odd;
(3-cycle)(3-cycle)(1-cycle) has order 3 and is even;
(3-cycle)(2-cycle)(2-cycle) has order 6 and is even;
(3-cycle)(2-cycle)(1-cycle)(1-cycle) has order 6 and is odd;
(3-cycle)(1-cycle)(1-cycle)(1-cycle) has order 3 and is even;
(2-cycle)(2-cycle)(2-cycle)(1-cycle) has order 2 and is odd;
(2-cycle)(2-cycle)(1-cycle)(1-cycle)(1-cycle) has order 2 and is even;
(2-cycle)(1-cycle)(1-cycle)(1-cycle)(1-cycle)(1-cycle) has order 2 and
is odd. So, in S_7, the possible orders are 1, 2, 3, 4, 5, 6, 7, 10, 12;
for A_7 the possible orders are 1, 2, 3, 4, 5, 6, 7.

9. $(135) = (15)(13)$ even; $(1356) = (16)(15)(13)$ odd; $(13567) =$
$(17)(16)(15)(13)$ even; $(12)(134)(152) = (12)(14)(13)(12)(15)$ odd;
$(1243)(3521) = (13)(14)(12)(31)(32)(35)$ even.

11. An n-cycle is even when n is odd since we can write it as a product
of $n - 1$ 2-cycles by successively pairing up the first element of the
cycle with each of the other cycle elements starting from the last
element of the cycle and working towards the front. The same
process shows that when n is odd we get an even permutation.

13. An even number of two cycles followed by an even number of two
cycles gives an even number of two cycles in all. So the Finite
Subgroup Test is verified.

15. Suppose that α can be written as a product on m 2-cycles and β
can be written as product of n 2-cycles. Then juxtaposing these
2-cycles we can write $\alpha\beta$ as a product of $m + n$ 2-cycles. Now
observe that $m + n$ is even if and only if m and n are even or both
odd.

17.

$$\alpha^{-1} = \begin{bmatrix} 1 & 2 & 3 & 4 & 5 & 6 \\ 2 & 1 & 3 & 5 & 4 & 6 \end{bmatrix}$$

$$\beta\alpha = \begin{bmatrix} 1 & 2 & 3 & 4 & 5 & 6 \\ 1 & 6 & 2 & 3 & 4 & 5 \end{bmatrix}$$

$$\alpha\beta = \begin{bmatrix} 1 & 2 & 3 & 4 & 5 & 6 \\ 6 & 2 & 1 & 5 & 3 & 4 \end{bmatrix}$$

19. If all members of H are even we are done. So, suppose that H has
at least one odd permutation σ. For each odd permutation β in H

observe that $\sigma\beta$ is even and, by cancellation, different βs give different $\sigma\beta$s. Thus, there are at least as many even permutations are there are odd ones. Conversely, for each even permutation β in H observe that $\sigma\beta$ is odd and, by cancellation, different βs give different $\sigma\beta$s. Thus, there are at least as many odd permutations are there are even ones.

21. No, the identity is even and the product of two odd permutations is an even permutation.

23. $C(\alpha_3) = \{\alpha_1, \alpha_2, \alpha_3, \alpha_4\}$, $C(\alpha_{12}) = \{\alpha_1, \alpha_7, \alpha_{12}\}$,

25. An odd permutation of order 4 must be of the form $(a_1 a_2 a_3 a_4)$. There are 6 choices for a_1, 5 for a_4, 4 for a_3, and 3 for a_4. This gives $6 \cdot 5 \cdot 4 \cdot 3$ choices. But since for each of these choices the cycles $(a_1 a_2 a_3 a_4) = (a_2 a_3 a_4 a_1) = (a_3 a_4 a_1 a_2) = (a_4 a_3 a_2 a_1)$ give the same group element we must divide $6 \cdot 5 \cdot 4 \cdot 3$ by 4 to obtain 90.

27. Since $\beta^{28} = (\beta^4)^7 = \epsilon$, we know that $|\beta|$ divides 28. But $\beta^4 \neq \epsilon$ so $|\beta| \neq 1, 2$, or 4. If $|\beta| = 14$, then β written in disjoint cycle form would need at least one 7-cycle and one 2-cycle. But that requires at least 9 symbols and we have only 7. Likewise, $|\beta| = 28$ requires at least one 7-cycle and one 4-cycle. So, $|\beta| = 7$. Thus, $\beta = \beta^8 = (\beta^4)^2 = (2457136)$.

29. Observe that if we start with a 9-cycle $(a_1 a_2 a_3 a_4 a_5 a_6 a_7 a_8 a_9)$ and cube it we get $(a_1 a_4 a_7)(a_2 a_5 a_8)(a_3 a_6 a_9)$. So, we can let $\sigma = (124586739)$. But $(157)(283)(469) = (157)(469)(283) = (283)(157)(469)$ so these give us (142568793) and (214856379).

31. Let $\alpha, \beta \in \mathrm{stab}(a)$. Then $(\alpha\beta)(a) = \alpha(\beta(a)) = \alpha(a) = a$. Also, $\alpha(a) = a$ implies $\alpha^{-1}(\alpha(a)) = \alpha^{-1}(a)$ or $a = \alpha^{-1}(a)$.

33. Since $\alpha^m = (1,3,5,7,9)^m (2,4,6)^m (8,10)^m$ and the result is a 5-cycle we deduce that $(2,4,6)^m = \epsilon$ and $(8,10)^m = \epsilon$. So, 3 and 2 divide m. Since $(1,3,5,7,9)^m \neq \epsilon$ we know that 5 does not divide m. Thus, we can say that m is a multiple of 6 but not a multiple of 30.

35. An element of order 5 in A_6 must be a 5-cycle. There are $6 \cdot 5 \cdot 4 \cdot 3 \cdot 2 = 720$ ways to create a 5-cycle. But the same 5-cycle can be written in 5 ways so we must divide 720 by 5 to obtain 144.

37. 3, 7, 9

39. Let $\alpha = (123)$ and $\beta = (145)$.

41. Observe that (12) and (123) belong to S_n for all $n \geq 3$ and they do not commute.

43. Cycle decomposition shows any nonidentity element of A_5 is a 5-cycle, a 3-cycle, or a product of a pair of disjoint 2-cycles. Then, observe there are $(5 \cdot 4 \cdot 3 \cdot 2 \cdot 1)/5 = 24$ group elements of the form $(abcde)$, $(5 \cdot 4 \cdot 3)/3 = 20$ group elements of the form (abc) and $(5 \cdot 4 \cdot 3 \cdot 2)/8 = 15$ group elements of the form $(ab)(cd)$. In this last case we must divide by 8 because there are 8 ways to write the same group element $(ab)(cd) = (ba)(cd) = (ab)(dc) = (ba)(dc) = (cd)(ab) = (cd)(ba) = (dc)(ab) = (dc)(ba)$.

45. Any element from A_n is expressible as a product of an even number of 2-cycles. For each pair of 2-cycles there are two cases. One is that they share an element in common $(ab)(ac)$ and the other is that they are disjoint $(ab)(cd)$. Now observe that $(ab)(ac) = (abc)$ and $(ab)(cd) = (cbd)(acb)$.

47. That $a * \sigma(b) \neq b * (a)$ is done by examining all cases. To prove the general case, observe that $\sigma^i(a) * \sigma^{i+1}(b) \neq \sigma^i(b) * \sigma^{i+1}(a)$ can be written in the form $\sigma^i(a) * \sigma(\sigma^i(b)) \neq \sigma^i(b) * \sigma(\sigma^i(a))$, which is the case already done. If a transposition were not detected, then
$\sigma(a_1) * \cdots * \sigma^i(a_i) * \sigma^{i+1}(a_{i+1}) * \cdots * \sigma^n(a_n) = \sigma(a_1) * \cdots * \sigma^i(a_{i+1}) * \sigma^{i+1}(a_i) * \cdots * \sigma^n(a_n)$, which implies $\sigma^i(a_i) * \sigma^{i+1}(a_{i+1}) = \sigma^i(a_{i+1}) * \sigma^{i+1}(a_i)$.

49. By Theorem 5.4 it is enough to prove that every 2-cycle can be expressed as a product of elements of the form $(1k)$. To this end observe that if $a \neq 1, b \neq 1$, then $(ab) = (1a)(1b)(1a)$.

51. If α has odd order k and α is an odd permutation then $\epsilon = a^k$ would be odd.

53. The product of an element from $Z(A_4)$ of order 2 and an element of A_4 of order 3 would have order 6. The product of an element from $Z(A_4)$ of order 3 and an element of A_4 of order 2 would have order 6. But A_4 has no element of order 6.

55. Labeling the four tires 1, 2, 3, 4 in clockwise order starting with 1 being the tire in the upper left-hand corner, we may represent the four patterns as:

$\alpha = (1324)$ top left-hand pattern

$\beta = (1423)$ top right-hand pattern

$\gamma = (14)(23)$ bottom right-hand pattern

$\delta = (13)(24)$ bottom left-hand pattern

Notice that $\alpha^{-1} = \beta$ and that $\delta = \alpha^2 \gamma$. Thus, we need only find the smallest subgroup of S_4 containing α and γ. To this end, observe that the set $\{\varepsilon, \alpha, \alpha^2, \alpha^3, \gamma, \alpha\gamma, \alpha^2\gamma, \alpha^3\gamma\}$ is closed under multiplication on the left and right by both α and γ. (Here we have used the fact that $\gamma\alpha = \alpha^3\gamma$.) This implies that the set is closed under multiplication and is therefore a group. Since $\alpha\gamma \neq \gamma\alpha$, the subgroup is non-Abelian.

CHAPTER 6
Isomorphisms

1. Let $\phi(n) = 2n$. Then ϕ is onto since the even integer $2n$ is the image of n. ϕ is one-to-one since $2m = 2n$ implies that $m = n$. $\phi(m + n) = 2(m + n) = 2m + 2n$ so ϕ is operation preserving.

3. ϕ is onto since any positive real number r is the image of $\sqrt{r}$. ϕ is one-to-one since $\sqrt{a} = \sqrt{b}$ implies that $a = b$. Finally, $\phi(xy) = \sqrt{xy} = \sqrt{x}\sqrt{y} = \phi(x)\phi(y)$.

5. Define ϕ from $U(8)$ to $U(12)$ by $\phi(1) = 1$; $\phi(3) = 5$; $\phi(5) = 7$; $\phi(7) = 11$. To see that ϕ is operation preserving we observe that
$\phi(1a) = \phi(a) = \phi(a) \cdot 1 = \phi(a)\phi(1)$ for all a;
$\phi(3 \cdot 5) = \phi(7) = 11 = 5 \cdot 7 = \phi(3)\phi(5)$;
$\phi(3 \cdot 7) = \phi(5) = 7 = 5 \cdot 11 = \phi(3)\phi(7)$;
$\phi(5 \cdot 7) = \phi(3) = 5 = 7 \cdot 11 = \phi(5)\phi(7)$.

7. D_{12} has an element of order 12 and S_4 does not.

9. Since $T_e(x) = ex = x$ for all x, T_e is the identity. For the second part, observe that $T_g \circ (T_g)^{-1} = T_e = T_{gg^{-1}} = T_g \circ T_{g^{-1}}$ and cancel.

11. For any x in the group, we have $(\phi_g\phi_h)(x) = \phi_g(\phi_h(x)) = \phi_g(hxh^{-1}) = ghxh^{-1}g^{-1} = (gh)x(gh)^{-1} = \phi_{gh}(x)$.

13. ϕ_{R_0} and $\phi_{R_{90}}$ disagree on H; ϕ_{R_0} and ϕ_H disagree on R_{90}; ϕ_{R_0} and ϕ_D disagree on R_{90}; $\phi_{R_{90}}$ and ϕ_H disagree on R_{90}; $\phi_{R_{90}}$ and ϕ_D disagree on R_{90}; ϕ_H and ϕ_D disagree on D.

15. Let $\alpha \in \text{Aut}(G)$. We show that α^{-1} is operation preserving: $\alpha^{-1}(xy) = \alpha^{-1}(x)\alpha^{-1}(y)$ if and only if $\alpha(\alpha^{-1}(xy)) = \alpha(\alpha^{-1}(x)\alpha^{-1}(y))$. That is, if and only if $xy = \alpha(\alpha^{-1}(x))\alpha(\alpha^{-1}(y)) = xy$. So α^{-1} is operation preserving. That $\text{Inn}(G)$ is a group follows from the equation $\phi_g\phi_h = \phi_{gh}$.

17. That α is a one-to-one follows from the fact that r^{-1} exists module n. The operation preserving condition is Exercise 11 of Chapter 0.

19. The inverse of a one-to-one function is one-to-one. To see that ϕ^{-1} is operation-preserving, let a and b belong to $\overline{G}$. Then
$\phi^{-1}(ab) = \phi^{-1}(a)\phi^{-1}(b)$ if and only if
$ab = \phi(\phi^{-1}(a))\phi(\phi^{-1}(b)) = ab$. [We obtain the first equality by
applying ϕ to both sides of $\phi^{-1}(ab) = \phi^{-1}(a)\phi^{-1}(b)$]. Finally, let
$g \in G$. Then $\phi^{-1}(\phi(g)) = g$, so that ϕ^{-1} is onto.

21. $T_g(x) = T_g(y)$ if and only if $gx = gy$ or $x = y$. This shows that T_g is a one-to one function. Let $y \in G$. Then $T_g(g^{-1}y) = y$, so that T_g is onto.

23. $\phi((a + bi) + (c + di)) = \phi((a + c) + (b + d)i) = (a + c) - (b + d)i = (a - bi) + (c - di) = \phi(a + bi) + \phi(c + di)$. Also,
$\phi((a + bi)(c + di)) = \phi((ac - bd) + (ad + bc)i) = (ac - bd) - (ad + bc)i$
and $\phi(a + bi)\phi(c + di) = (a - bi)(c - di) = (ac - bd) - (ad + bc)i$.

25. First observe the Z is a cyclic group generated by 1. By property 2 of Theorem 6.3, it suffices to show that Q is not cyclic under addition. By way of contradiction suppose that $Q = \langle p/q \rangle$. But then $p/2q$ is a rational number that is not in $\langle p/q \rangle$.

27. The notation itself suggests that

$$\phi(a + bi) = \begin{bmatrix} a & -b \\ b & a \end{bmatrix}$$

is the appropriate isomorphism. To verify this note that

$$\phi((a + bi) + (c + di)) = \begin{bmatrix} a + c & -(b + d) \\ (b + d) & a + c \end{bmatrix} =$$

$$\begin{bmatrix} a & -b \\ b & a \end{bmatrix} + \begin{bmatrix} c & -d \\ d & c \end{bmatrix} = \phi(a + bi) + \phi(c + di).$$

Also, $\phi((a + bi)(c + di)) = \phi((ac - bd) + (ad + bc)i) =$

$$\begin{bmatrix} (ac - bd) & -(ad + bc)) \\ (ad + bc) & ac - bd) \end{bmatrix} = \begin{bmatrix} a & -b \\ b & a \end{bmatrix}\begin{bmatrix} c & -d \\ d & c \end{bmatrix} =$$

$$\phi(a + bi)\phi(c + di).$$

29. Yes, by Cayley's Theorem.

31. Observe that $\phi_g(y) = gyg^{-1}$ and
$\phi_{zg}(y) = zgy(zg)^{-1} = zgyg^{-1}z^{-1} = gyg^{-1}$, since $z \in Z(G)$. So,
$\phi_g = \phi_{zg}$.

33. $\phi_g = \phi_h$ implies $gxg^{-1} = hxh^{-1}$ for all x. This implies
$h^{-1}gx(h^{-1}g)^{-1}) = x$, and therefore $h^{-1}g \in Z(G)$.

35. Say $|a| = n$. Then $\phi_a^n(x) = a^n x a^{-n} = x$, so that ϕ_a^n is the identity.
For example, take $a = R_{90}$ in D_4. Then $|\phi_a| = 2$.

37. Say ϕ is an isomorphism from Q to $\mathbf{R}^+$ and ϕ takes 1 to a. It
follows that the integer r maps to a^r. Then
$a = \phi(1) = \phi(s\frac{1}{s}) = \phi(\frac{1}{s} + \cdots + \frac{1}{s}) = \phi(\frac{1}{s}))^s$ and therefore
$a^{\frac{1}{s}} = \phi(\frac{1}{s})$. Thus, the rational r/s maps to $a^{r/s}$. But $a^{r/s} \neq a^\pi$ for
any rational number r/s.

39. $(R_0 R_{90} R_{180} R_{270})(H D'V D)$.

41. The mapping $\phi(x) = x^2$ is one-to-one from Q^+ to Q^+ since $a^2 = b^2$
implies $a = b$ when both a and b are positive. Moreover,
$\phi(ab) = \phi(a)\phi(b)$ for all a and b. However, ϕ is not onto since there
is no rational whose square is 2. So, the image of ϕ is a proper
subgroup of Q^+.

43. Suppose that ϕ is an automorphism of R^* and a is positive. Then
$\phi(a) = \phi(\sqrt{a}\sqrt{a}) = \phi(\sqrt{a})\phi(\sqrt{a}) = \phi(\sqrt{a})^2 > 0$. Now suppose that
a is negative but $\phi(a) = b$ is positive. Then by the case we just did
$a = \phi^{-1}(\phi(a)) = \phi^{-1}(b)$ is positive. This is a contradiction.

CHAPTER 7
Cosets and Lagrange's Theorem

1. $H = \{\alpha_1, \alpha_2, \alpha_3, \alpha_4\}$, $\alpha_5 H = \{\alpha_5, \alpha_8, \alpha_6, \alpha_7\}$,
   $\alpha_9 H = \{\alpha_9, \alpha_{11}, \alpha_{12}, \alpha_{10}\}$.

3. $H, 1 + H, 2 + H$. To see that there are no others notice that for any
   integer n we can write $n = 3q + r$ where $0 \leq r < 3$. So,
   $n + H = r + 3q + H = r + H$, where $r = 0, 1$ or 2.

5. **a.** $11 + H = 17 + H$ because $17 - 11 = 6$ is in H;

   b. $-1 + H = 5 + H$ because $5 - (-1) = 6$ is in H;

   c. $7 + H \neq 23 + H$ because $23 - 7 = 16$ is not in H.

7. Since $8/2 = 4$, there are four cosets. Let $H = \{1, 11\}$. The cosets
   are $H, 7H, 13H, 19H$.

9. Since $|a^4| = 15$, there are two cosets: $\langle a^4 \rangle$ and $a \langle a^4 \rangle$.

11. Suppose that $h \in H$ and $h < 0$. Then $h\mathbf{R}^+ \subseteq hH = H$. But $h\mathbf{R}^+$ is
    the set of all negative real numbers. Thus $H = \mathbf{R}^*$.

13. By Lagrange's Theorem the possible orders are 1, 2, 3, 4, 5, 6, 10,
    12, 15, 20, 30, 60.

15. By Lagrange's Theorem, the only possible orders for the subgroups
    are 1, p and q. By Corollary 3 of Lagrange's Theorem, groups of
    prime order are cyclic. The subgroup of order 1 is $\langle e \rangle$.

17. By Exercise 16 we have $5^6 \bmod 7 = 1$. So, using mod 7 we have
    $5^{15} = 5^6 \cdot 5^6 \cdot 5^2 \cdot 5 = 1 \cdot 1 \cdot 4 \cdot 5 = 6$; $7^{13} \bmod 11 = 2$.

19. By Theorem 0.2 there are integers s and t such that $1 = ms + nt$.
    Then $g = g^{ms+nt} = (g^m)^s (g^n)^t = e^s e^t = e$.

21. Since G has odd order, no element can have order 2. Thus, for each
    $x \neq e$, we know that $x \neq x^{-1}$. So, we can write the product of all
    elements in the form $ea_1 a_1^{-1} a_2 a_2^{-1} \cdots a_n a_n^{-1} = e$.

23. Let H be the subgroup of order p and K be the subgroup of order q. Then $H \cup K$ has $p + q - 1 < pq$ elements. Let a be any element in G that is not in $H \cup K$. By Lagrange's Theorem, $|a| = p, q,$ or pq. But $|a| \neq p$, for if so, then $\langle a \rangle = H$. Similarly, $|a| \neq q$.

25. The possible orders are 1, 3, 11, 33. If $|x| = 33$, then $|x^{11}| = 3$ so we may assume that there is no element of order 33. By the Corollary of Theorem 4.4, the number of elements of order 11 is a multiple of 10 so they account for 0, 10, 20, or 30 elements of the group. The identity accounts for one more. So, at most we have accounted for 31 elements. By Corollary 2 of Lagrange's Theorem, the elements unaccounted for have order 3.

27. Observe that $|G : H| = |G|/|H|, |G : K| = |G|/|K|,$ $|K : H| = |K|/|H|.$ So, $|G : K||K : H| = |G|/|H| = |G : H|.$

29. Certainly, $a \in \mathrm{orb}_G(a)$. Now suppose $c \in \mathrm{orb}_G(a) \cap \mathrm{orb}_G(b)$. Then $c = \alpha(a)$ and $c = \beta(b)$ for some α and β, and therefore $(\beta^{-1}\alpha)(a) = \beta^{-1}(\alpha(a)) = \beta^{-1}(c) = b$. So, if $x \in \mathrm{orb}_G(b)$, then $x = \gamma(b) = \gamma((\beta^{-1}\alpha)(a)) = (\gamma\beta^{-1}\alpha)(a)$. This proves $\mathrm{orb}_G(b) \subseteq \mathrm{orb}_G(a)$. By symmetry, $\mathrm{orb}_G(a) \subseteq \mathrm{orb}_G(b)$.

31.   **a.** $\mathrm{stab}_G(1) = \{(1), (24)(56)\}; \mathrm{orb}_G(1) = \{1, 2, 3, 4\}$

   b. $\mathrm{stab}_G(3) = \{(1), (24)(56)\}; \mathrm{orb}_G(3) = \{3, 4, 1, 2\}$

   c. $\mathrm{stab}_G(5) = \{(1), (12)(34), (13)(24), (14)(23)\}; \mathrm{orb}_G(5) = \{5, 6\}$

33. Suppose that $|Z(G)| = p^{n-1}$ and let a be an element of G not in $|Z(G)|$. Then $C(a)$ contains both a and $Z(G)$. By Lagrange's Theorem we must have $C(a) = G$. But then $a \in Z(G)$.

35. By Lagrange's Theorem the order must be divisible by 1, 2, ..., 10. So, the order must be divisible by the least common multiple of these which is 2520.

37. Suppose $x^2 = y^2$ and let $|G| = 2k + 1$. Then $x = xe = xx^{2k+1} = x^{2k+2} = (x^2)^{k+1} = (y^2)^{k+1} = y^{2k+2} = yy^{2k+1} = ye = y.$

39. Suppose that $B \in G$ and det $B = 2$. Then det $A^{-1}B = 1$, so that $A^{-1}B \in H$ and therefore $B \in AH$. Conversely, for any $Ah \in AH$ we have det $Ah = ($ det $A)($det $h) = 2 \cdot 1 = 2.$

41. It is the set of all permutations that carry face 2 to face 1.

43. If $aH = bH$, then $b^{-1}a \in H$. So det $(b^{-1}a) = (\det b^{-1})(\det a) =$ $(\det b^{-1})(\det a) = (\det b)^{-1}(\det a) = 1$. Thus det $a = \det b$. Conversely, we can read this argument backwards to get that det a $= \det b$ implies $aH = bH$.

45. Since the order of G is divisible by both 10 and 25 it must be divisible by 50. But the only number less than 100 that is divisible by 50 is 50.

47. For any positive integer n let $\omega_n = \cos\left(\frac{2\pi}{n}\right) + i\sin\left(\frac{2\pi}{n}\right)$. The finite subgroups of C^* are those of the form $\langle\omega_n\rangle$. To verify this, let H denote any finite subgroup of C^* of order n. Then every element of H is a solution to $x^n = 1$. But the solution set of $x^n = 1$ in C^* is $\langle\omega_n\rangle$.

CHAPTER 8
External Direct Products

1. Closure and associativity in the product follows from the closure and associativity in each component. The identity in the product is the n-tuple with the identity in each component. The inverse of $(g_1, g_2, \ldots, g_n)$ is $(g_1^{-1}, g_2^{-1}, \ldots, g_n^{-1})$.

3. The mapping $\phi(g) = (g, e_H)$ is an isomorphism from G to $G \oplus \{e_H\}$. To verify that ϕ is one-to-one, we note that $\phi(g) = \phi(g')$ implies $(g, e_H) = (g', e_H)$ which means that $g = g'$. The element $(g, e_H) \in G \oplus \{e_H\}$ is the image of g. Finally, $\phi((g, e_H)(g', e_H)) = \phi((gg', e_H e_H)) = \phi((gg', e_H)) = gg' = \phi((g, e_H)\phi((g', e_H))$. A similar argument shows that $\phi(h) = (e_G, h)$ is an isomorphism from H onto $\{e_G\} \oplus H$.

5. To show that $Z \oplus Z$ is not cyclic note that $(a, b+1) \notin \langle (a, b) \rangle$.

7. Define a mapping from $G_1 \oplus G_2$ to $G_2 \oplus G_1$ by $\phi(g_1, g_2) = (g_2, g_1)$. To verify that ϕ is one-to-one, we note that $\phi((g_1, g_2)) = \phi((g_1', g_2'))$ implies $(g_2, g_1) = (g_2', g_1')$. From this we obtain that $g_1 = g_1'$ and $g_2 = g_2'$. The element (g_2, g_1) is the image on (g_1, g_2) so ϕ is onto. Finally, $\phi((g_1, g_2)(g_1', g_2')) = \phi((g_1 g_1', g_2 g_2')) = (g_2 g_2', g_1 g_1') = (g_2, g_1)(g_2', g_1') = \phi((g_1, g_2))\phi((g_1', g_2'))$

9. Yes, By Theorem 8.2.

11. In both $Z_4 \oplus Z_4$ and $Z_{8000000} \oplus Z_{400000}$, $|(a, b)| = 4$ if and only if $|a| = 4$ and $|b| = 1, 2$ or 4 or if $|b| = 4$ and $|a| = 1$ or 2 (we have already counted the case that $|a| = 4$). For the first case, we have $\phi(4) = 2$ choices for a and $\phi(4) = \phi(2) + \phi(1) = 4$ choices for b to give us 8 in all. For the second case, we have $\phi(4) = 2$ choices for b and $\phi(2) + \phi(1) = 2$ choices for b. This gives us a total of 12.

In the general case observe that by Theorem 4.4 that as long as d divides n the number of elements of order d in a cyclic group depends only on d.

13. Define a mapping ϕ from $\mathbf{C}$ to $\mathbf{R} \oplus \mathbf{R}$ by $\phi(a + bi) = (a, b)$. To verify that ϕ is one-to-one note that $\phi(a + bi) = \phi(a' + b'i)$ implies that $(a, b) = (a', b')$. So, $a = a'$ and $b = b'$ and therefore $a + bi = a' + b'i$. The element (a, b) in $\mathbf{R} \oplus \mathbf{R}$ is the image of $a + bi$ so ϕ is onto. Finally,
$\phi((a + bi) + (a' + b'i)) = \phi((a + a') + (b + b')i) = (a + a', b + b') = (a, b) + (a', b') = \phi(a + bi) + \phi(a' + b'i)$.

15. By Exercise 3 in this chapter G is isomorphic to $G \oplus \{e_H\}$ and H is isomorphic to $\{e_G\} \oplus H$. Since subgroups of cyclic groups are cyclic, we know that $G \oplus \{e_H\}$ and $\{e_G\} \oplus H$ are cyclic.

17. $|(a, b, c)| = \text{lcm}\{|a|, |b|, |c|\} = 3$, unless $a = b = c = e$.

19. Define a mapping ϕ from M to N by $\phi\left(\begin{bmatrix} a & b \\ c & d \end{bmatrix}\right) = (a, b, c, d)$.
To verify that ϕ is one-to-one we note that
$\phi\left(\begin{bmatrix} a & b \\ c & d \end{bmatrix}\right) = \phi\left(\begin{bmatrix} a' & b' \\ c' & d' \end{bmatrix}\right)$ implies $(a, b, c, d) = (a', b', c', d')$.
Thus $a = a', b = b', c = c'$, and $d = d'$. This proves that ϕ is one-to-one. The element (a, b, c, d) is the image of $\begin{bmatrix} a & b \\ c & d \end{bmatrix}$ so ϕ is onto. Finally, $\phi\left(\begin{bmatrix} a & b \\ c & d \end{bmatrix} + \begin{bmatrix} a' & b' \\ c' & d' \end{bmatrix}\right) =$
$\phi\left(\begin{bmatrix} a + a' & b + b' \\ c + c' & d + d' \end{bmatrix}\right) = (a + a', b + b', c + c', d + d') =$
$(a, b, c, d) + (a', b', c', d') = \phi\left(\begin{bmatrix} a & b \\ c & d \end{bmatrix}\right) + \phi\left(\begin{bmatrix} a' & b' \\ c' & d' \end{bmatrix}\right)$.
Let $\mathbf{R}^k$ denote $\mathbf{R} \oplus \mathbf{R} \oplus \cdots \oplus \mathbf{R}$ (k factors). Then the group of $m \times n$ matrices under addition is isomorphic to R^{mn}.

21. Since $(g, g)(h, h)^{-1} = (gh^{-1}, gh^{-1})$, H is a subgroup. When $G = \mathbf{R}$, $G \oplus G$ is the plane and H is the line $y = x$.

23. $\langle(3, 0)\rangle, \langle(3, 1)\rangle, \langle(3, 2)\rangle, \langle(0, 1)\rangle$

25. $|(1, 1)| = \text{lcm}\{30, 20\} = 60$.

27. $\langle 400 \rangle \oplus \langle 50 \rangle = \{0, 400\} \oplus \{0, 50, 100, 150\}$

29. In $\mathbf{R}^* \oplus \mathbf{R}^*$ $(1, -1)$, $(-1, 1)$ and $(-1, -1)$ have order 2 whereas in $\mathbf{C}^*$ the only element of order 2 is -1. But isomorphisms preserve order.

31. Define the mapping from G to $Z \oplus Z$ by $\phi(3^m 6^n) = (m, n)$. To verify that ϕ is one-to-one note that $\phi(3^m 6^n) = \phi(3^s 6^t)$ implies that $(m, n) = (s, t)$, which in turn implies that $m = s$ and $n = t$. So, $3^m 6^n = 3^s 6^t$. The element (m, n) is the image of $3^m 6^n$ so ϕ is onto. Finally, $\phi((3^m 6^n)(3^s 6^t)) = \phi(3^{m+s} 6^{n+t}) = (m + s, n + t) = (m, n) + (s, t) = \phi(3^m 6^n)\phi(3^s 6^t)$ shows that ϕ is operation preserving.

33. D_{24} has an element of order 24, whereas $D_3 \oplus D_4$ does not.

35. Each cyclic subgroup of order 6 has two elements of order 6. So, the 24 elements of order 6 yield 12 cyclic subgroups of order 6.

37. $\text{Aut}(U(25)) \approx \text{Aut}(Z_{20}) \approx U(20) \approx U(4) \oplus U(5) \approx Z_2 \oplus Z_4$.

39. In each position we must have an element of order 1 or 2 except for the case that every position has the identity. So, there are $2^k - 1$ choices. For the second question, we must use the identity in every position for which the order of the group is odd. So, there are $2^t - 1$ elements of order 2 where t is the number of the $n_1, n_2, \ldots, n_k$ that are even.

41. No. $Z_{10} \oplus Z_{12} \oplus Z_6$ has 7 elements of order 2 whereas $Z_{15} \oplus Z_4 \oplus Z_{12}$ has only 3.

43. Using the fact that an isomorphism from Z_{12} is determined by the image of 1 and the fact that a generator must map to a generator, we determine that there are 4 isomorphisms.

45. Since $a \in Z_m$ and $b \in Z_n$, we know that $|a|$ divides m and $|b|$ divides n. So, $|(a, b)| = \text{lcm}(|a|, |b|)$ divides $\text{lcm}(m, n)$.

47. Up to isomorphism, Z is the only infinite cyclic group and it has 1 and -1 as its only generators. The number of generators of Z_m is $|U(m)|$ so we must determine those m such that $|U(m)| = 2$. First consider the case where $m = p^n$, where p is a prime. Then the number of generators is $p^{n-1}(p - 1)$. So, if $p > 3$ we will have more than 2 generators. When $p = 3$ we must have $n = 1$. Finally, $|U(2^n)| = 2^{n-1} = 2$ only when $n = 2$. This gives us Z_3 and Z_4.

When $m = p_1 p_2 \cdots p_k$, where the p's are distinct primes we have $|U(m)| = |U(p_1)||U(p_2)| \cdots |U(p_k))|$. As before no prime can be greater than 3. So, the only case is $m = 2 \cdot 3 = 6$.

49. Each subgroup of order p consists of the identity and $p - 1$ elements of order p. So, we count the number of elements of order p and divide by $p - 1$. In $Z_p \oplus Z_p$ every nonidentity element has order p so there are $(p^2 - 1)/(p - 1) = p + 1$ subgroups of order p.

51. $U(165) \approx U(15) \oplus U(11) \approx U(5) \oplus U(33) \approx U(3) \oplus U(55) \approx U(3) \oplus U(5) \oplus U(11)$.

53. We use the fact that $\text{Aut}(Z_{720}) \approx U(720) \approx Z_2 \oplus Z_4 \oplus Z_6 \oplus Z_4$. In order for (a, b, c, d) to have order 6 we must have $|c| = 6$ and a, b, and d have orders 1 or 2. So we have 2 choices for each of a, b, c, and d. This gives 16 in all.

55. $U(900) \approx Z_2 \oplus Z \oplus Z_{20}$ so the element of largest order is the lcm(2, 6, 20) = 60.

57. Observe that $U(55) \approx U(5) \oplus U(11) \approx Z_4 \oplus Z_{10}$ and $U(75) \approx U(3) \oplus U(25) \approx Z_2 \oplus Z_{20} \approx Z_2 \oplus Z_5 \oplus Z_4 \approx Z_{10} \oplus Z_4 \approx Z_4 \oplus Z10$.

59. By the Finite Subgroup Test, the set is closed since $a^2 b^2 = (ab)^2$. The set is proper because $1^2 = (n - 1)^2$.

61. We need to find relatively prime m and n so that both $|U(m)|$ and $|U(n)|$ are divisible by 5. Since for every odd prime p we have $|U(p^k)| = p^{k-1}(p - 1)$, we can choose $m = 5^2$ and $n = 11$. Then $U(mn) = U(275) \approx U(25) \oplus U(11) \approx Z_{20} \oplus Z_{10}$.

63. This is the same argument as in Exercise 61 with 5 replaced by 3. So, $m = 9$ and $n = 7$ gives us $U(63) \approx U(9) \oplus U(7) \approx Z_6 \oplus Z_6$, which contains $\langle 2 \rangle \oplus \langle 2 \rangle \approx Z_3 \oplus Z_3$.

65. We need a prime-power p^k so that $p^{k-1}(p - 1)$ is divisible by 14. Obviously, 7^2 is one choice.

67. This time we need p^s and q^t, where p and q are distinct primes and both $p^{s-1}(p - 1)$ and $q^{t-1}(q - 1)$ to be divisible by 4. One such choice is $p = 5$ and $q = 13$. The $U(65) \approx U(5) \oplus U(13) \approx Z_4 \oplus Z_{12}$, which contains $\langle 1 \rangle \oplus \langle 3 \rangle \approx Z_4 \oplus Z_4$.

69. Since $5 \cdot 29 = 1 \mod 36$, we have that $s = 29$. So, we need to compute $34^{29} \mod 2701$. The result is 1415, which converts to NO.

SUPPLEMENTARY EXERCISES FOR CHAPTERS 5-8

1. Consider the finite and infinite cases separately. In the finite case, note that $|H| = |\phi(H)|$. Now use Theorem 4.3. For the infinite case, use Exercise 2 of Chapter 6.

3. Observe that $\phi(x^{-1}y^{-1}xy) = (\phi(x))^{-1}(\phi(y))^{-1}(\phi(x))(\phi(y))$, so ϕ carries the generators of G to the generators of G'.

5. All nonidentity elements of G and H have order 3. Since G is Abelian and H is non-Abelian, $G \not\approx H$.

7. Certainly the set HK has $|H||K|$ symbols. However, not all symbols need represent distinct group elements. That is, we may have $hk = h'k'$ although $h \neq h'$ and $k \neq k'$. We must determine the extent to which this happens. For every t in $H \cap K$ $hk = (ht)(t^{-1}k)$, so each group element in HK is represented by at least $|H \cap K|$ products in HK. But $hk = h'k'$ implies $t = h^{-1}h' = k(k')^{-1} \in H \cap K$, so that $h' = ht$ and $k' = t^{-1}k$. Thus each element in HK is represented by exactly $|H \cap K|$ products. So, $|HK| = |H||K|/|H \cap K|$.

9. For $U(n)$ to have exponent 2 every nonidentity element must have order 2. If a prime p divides n and $p > 3$, then $U(n)$ has an element of order $p - 1$ so the only primes that can divide n are 2 and 3. Since $U(3^m)$ has an element of order 3^{m-1}, 3^2 cannot divide n. Since $U(2^m)$ has an element of order 2^{m-2} when $m > 3$ and $U(2)$ is the identity we know the only possible powers of 2 that can divide n are 2, 4 and 8. Thus, we have shown that n must have the form $3^s 2^t$ where $s = 0$ or 1 and $t = 0, 1, 2$, or 3. This gives us $n = 1, 2, 4, 8, 3, 6, 12$, or 24. But $U(1)$ and $U(2)$ have exponent 1 so we are left with $U(n)$, where $n = 4, 8, 3, 6, 12, 24$.

11. Observe that $a + bi = \sqrt{a^2 + b^2} \left(\dfrac{a}{\sqrt{a^2 + b^2}} + \dfrac{b}{\sqrt{a^2 + b^2}}i \right)$, $\sqrt{a^2 + b^2}$ belongs to $\mathbf{R}^+$ and $\dfrac{a}{\sqrt{a^2+b^2}} + \dfrac{b}{\sqrt{a^2+b^2}}i \in T$ so every element can be expressed in the desired form.

Now suppose that $r_1 z_1 = r_2 z_2$, where r_1 and r_2 belong to $\mathbf{R}^+$ and z_1 and z_2 belong to T. Then $r_1 r_2^{-1} = z_2 z_1^{-1} \in T$. Thus, $(r_1 r_2^{-1})^2 = 1$. Since $r_1 r_2^{-1}$ is positive, we have $r_1 r_2^{-1} = 1$. So, $r_1 = r_2$ and $z_1 = z_2$.

13. Suppose $\phi : Q \to R$ is an isomorphism. Let $\phi(1) = x_0$. By property 2 of Theorem 6.2, for any nonzero integer b, we have $x_0 = \phi(b\frac{1}{b}) = b\phi(\frac{1}{b})$ so that $\phi(\frac{1}{b}) = \frac{1}{b}x_0$. Also, $\phi(\frac{a}{b}) = a\phi(\frac{1}{b}) = a\frac{1}{b}x_0 = \frac{a}{b}x_0$. So, if x_0 is irrational, then 1 is not in the image and if x_0 is rational the $\sqrt{2}$ is not in the image.

15. In Q, the equation $2x = a$ has a solution for all a. The corresponding equation $x^2 = b$ in Q^+ does not have a solution for $b = 2$.

17. Suppose $x^{p-2} = 1$. Since $|U(p)| = p - 1$, we have that $x^{p-1} = 1$ for all $x \in U(p)$. So, cancellation, $x = 1$.

19. $\langle 3 \rangle \oplus \langle 4 \rangle$

21. Z_{18}, $Z_2 \oplus Z_3 \oplus Z_3$, D_9, $D_3 \oplus Z_3$. To see that these are not isomorphic note that Z_{18} is cyclic and the others are not, $Z_2 \oplus Z_3 \oplus Z_3$ is Abelian and the last two are not, D_9 has an element of order 9 and $D_3 \oplus Z_3$ does not.

23. Say $\alpha = a_1 a_2 \cdots a_n$ and $\beta = b_1 b_2 \ldots b_m$, where the a's and b's are cycles. Then $\alpha\beta^{-1} = a_1 a_2 \cdots a_n b_m^{-1} \cdots b_1^{-1}$ is expressed as a product of a finite number of cycles.

25. $D_{11} \oplus Z_3$ has 11 elements of order 2 whereas $D_3 \oplus Z_{11}$ has only 3.

27. D_{33} has 33 elements of order 2 whereas $D_3 \oplus Z_{11}$ has only 3.

29. If $|\text{Inn}(G)| = 1$, then for all a in G, $x = \phi_a(x) = axa^{-1}$, so that $xa = ax$. Conversely, if G is Abelian, ϕ_a is the identity.

31. $U_{50}(450) \approx U(9) \approx Z_6$

33. $(4, 10)$

35. D_{12} has 13 elements of order 2 whereas $Z_3 \oplus D_4$ has only 5.

37. $20; (8, 7, (3251))$

39. Z_8, $Z_4 \oplus Z_2$, $Z_2 \oplus Z_2 \oplus Z_2$, D_4, and the quaternions. To verify that not two of these are isomorphic note that Z_8 is cyclic and the others are not; $Z_4 \oplus Z_2$ is Abelian and has an element of order 4 whereas $Z_2 \oplus Z_2 \oplus Z_2$ has no element of order 2 and D_4 and the quarterions are non-Abelian; $Z_2 \oplus Z_2 \oplus Z_2$ is Abelian and D_4 and the quaterions are not; D_4 has 5 elements of order 2 whereas the quarterions has only 1.

41. $(12)(34)(56789)$.

43. $\{(1), (345), (354), (12), (12)(345), (12)(354)\}$.

45. Since $|\beta^2| = 15$, we know that $|\beta| = 15$ or 30. But A_9 has no element of order 30, so $|\beta| = 15$. Then
$\beta = \beta^{16} = (\beta^2)^8 = (17395)(286)$

47. Say the points in H lie on the line $y = mx$. Then
$(a, b) + H = \{(a + x, b + mx) \mid x \in \mathbf{R}\}$. This set is the line
$y - b = m(x - a)$.

49. $aH = bH$ implies $a^{-1}b \in H$. So $(a^{-1}b)^{-1} = b^{-1}a \in H$. Thus, $Hb^{-1}a = H$ or $Hb^{-1} = Ha^{-1}$. These steps are reversible.

51. Let β have order 2. In disjoint cycle form, β is a product of transpositions, and since n is odd there must be some i missing from this product. Thus, $\beta(i) = i$. Pick j so that $\beta(j) \neq j$. Since σ is an n-cycle, some power of σ, say σ^t, takes i to j. If β commutes with σ, it commutes with σ^t as well. Then
$(\sigma^t \beta)(i) = \sigma^t(\beta(i)) = \sigma^t(i) = j$, whereas
$(\beta \sigma^t)(i) = \beta(\sigma^t(i)) = \beta(j) \neq j$. This proves that $\sigma^t \beta \neq \beta \sigma^t$.

CHAPTER 9
Normal Subgroups and Factor Groups

1. No, $(13)(12)(13)^{-1} = (23)$ is not in H.

3. Say $i < j$ and $h \in H_i \cap H_j$. Then $h \in H_1 H_2 \cdots H_{j-1} \cap H_j = \{e\}$.

5. H contains the identity so H is not empty. Let $A, B \in H$. Then $\det(AB^{-1}) = (\det A)(\det B)^{-1} \in K$. This proves that H is a subgroup. Also, for $A \in H$ and $B \in G$ we have $\det(BAB^{-1}) = (\det B)(\det A)(\det B)^{-1} = \det A \in K$ so $BAB^{-1} \in H$.

7. Let $x \in G$. If $x \in H$, then $xH = H = Hx$. If $x \notin H$, then xH is the set of elements in G, not in H and Hx is also the elements in G, not in H.

9. In G/H the element $(1,1)H$ has order 4, so $G/H \approx Z_4$. In G/K the elements $(1,1)K, (3,3)K$, and $(2,3)K$ have order 2, so $G/K \approx Z_2 \oplus Z_2$.

11. No. In S_3, let $H = \langle(123)\rangle$. Then H and S_3/H are Abelian but S_3 is not.

13. This follows directly from $(ab)h = a(bh)$ for all $h \in H$.

15. Since $(4U_5(105))^2 = 16U_5(105) = U_5(105)$, $|4U_5(105)| = 2$.

17. $H = \{0 + \langle20\rangle, 4 + \langle20\rangle, 8 + \langle20\rangle, 12 + \langle20\rangle, 16 + \langle20\rangle\}$.
$G/H = \{0 + \langle20\rangle + H, 1 + \langle20\rangle + H, 2 + \langle20\rangle + H, 3 + \langle20\rangle + H\}$

19. $40/10 = 4$

21. By Theorem 9.5 the group has an element a of order 3 and an element b of order 11. Then $|ab| = 33$.

23. Since $(1,1) + \langle(4,2)\rangle$ has infinite order, $(Z \oplus Z)/\langle(4,2)\rangle$ is infinite. It is not cyclic because $(6,3) + \langle(4,2)\rangle$ has order 2 and any infinite cyclic group is isomorphic to Z, which has no elements of finite order other than the identity.

25. Since the element $|3H|$ of G/H has order 8, $G/H \approx Z_8$.

27. Since H and K have order 2 they are both isomorphic to Z_2 and therefore isomorphic to each other. Since $|G/H| = 4$ and $|3H| = 4$ we know that $G/H \approx Z_4$. On the other hand, direct calculations show that each of the three nonidentity elements in G/K has order 2, so $G/K \approx Z_2 \oplus Z_2$.

29. Since det $(AB^{-1}) = (\det A)(\det B)^{-1}$, H is a subgroup. Since det $(CAC^{-1}) = (\det C)(\det A)(\det C)^{-1} = \det A$, H is normal.

31. Certainly, every nonzero real number is of the form $\pm r$, where r is a positive real number. Real numbers commute, and $\mathbf{R}^+ \cap \{1, -1\} = \{1\}$.

33. In the general case that $G = HK$ there is no relationship. If $G = H \times K$, then $|g| = \text{lcm}(|h|, |k|)$ provided the $|h|$ and $|k|$ are finite. If $|h|$ or $|k|$ is infinite, so is $|g|$.

35. For the first question, note that $\langle 3 \rangle \cap \langle 6 \rangle = \{1\}$ and $\langle 10 \rangle \cap \langle 3 \rangle \langle 6 \rangle = \{1\}$. For second question, observe that $12 = 3^{-1}6^2$ so $(\langle 3 \rangle) \times \langle 12 \rangle \neq \emptyset$.

37. Say $|g| = n$. Then $(gH)^n = g^n H = eH = H$. Now use Corollary 2 to Theorem 4.1.

39. Let $x \in C(H)$ and $g \in G$. We must show that $gxg^{-1} \in C(H)$. That is, for any h in H, $gxg^{-1}h = hgxg^{-1}$. Note that in the expression $(gxg^{-1})h(gxg^{-1})^{-1} = gxg^{-1}hgx^{-1}g^{-1}$ the terms x and x^{-1} cancel since $g^{-1}hg \in H$ and x commutes with every element of H. Then we have $(gxg^{-1})h(gxg^{-1})^{-1} = gxg^{-1}hgx^{-1}g^{-1} = gg^{-1}hgg^{-1} = h$. So, $gxg^{-1} \in C(H)$.

41. Take $G = Z_6$, $H = \{0, 3\}$, $a = 1$, and $b = 4$.

43. By Lagrange's Theorem, $|Z(G)| = 1, p, p^2$, or p^3. By assumption, $|Z(G)| \neq 1$ or p^3 (for then G would be Abelian). So, $|Z(G)| = p$ or p^2. However, the "G/Z" Theorem (Theorem 9.3) rules out the latter case.

45. If H is normal in G, then $xNhN(xN)^{-1} = xhx^{-1}N \in H/N$, so H/N is normal in G/N. Now assume H/N is normal in G/N. Then

$xhx^{-1}N = xNhN(xN)^{-1} \in H/N$. Thus $xhx^{-1}N = h'N$ for some $h' \in H$ and therefore $xhx^{-1} \in H$. So, $xhx^{-1} = h'n$ for some $n \in N$.

47. Say H has an index n. Then $(\mathbf{R}^*)^n = \{x^n \mid x \in \mathbf{R}^*\} \subseteq H$. If n is odd, then $(\mathbf{R}^*)^n = \mathbf{R}^*$; if n is even, then $(\mathbf{R}^*)^n = \mathbf{R}^+$. So, $H = \mathbf{R}^*$ or $H = \mathbf{R}^+$.

49. By Exercise 7, we know that K is normal in L and L is normal in D_4. But $VK = \{V, R_{270}\}$ whereas $KV = \{V, R_{90}\}$. So, K is not normal in D_4.

51. Suppose n_1h_1 and $n_2h_2 \in NH$. Then $n_1h_1n_2h_2 = n_1n'h_1h_2 \in NH$. Also $(n_1h_1)^{-1} = h_1^{-1}n_1^{-1} = nh_1^{-1} \in NH$. For the second part in S_3 take $N = \{(1), (12)\}$ and $M = \{(1), (13)\}$.

53. Let $H = \langle a^k \rangle$ be any subgroup of $N = \langle a \rangle$. Let $x \in G$ and let $(a^k)^m \in H$. We must show that $x(a^k)^m x^{-1} \in H$. Note that $x(a^{km}x^{-1} = (xax^{-1})^{km} = (a^r)^{km} = (a^k)^{rm} \in \langle a^k \rangle$. (Here we used the normality of N to replace xax^{-1} by a^r.)

55. $\gcd(|x|, |G|/|H|) = 1$ implies $\gcd(|xH|, |G/H|) = 1$. But $|xH|$ divides G/H. Thus $|xH| = 1$ and therefore $xH = H$.

57. By Corollary 4 of Theorem 7.1, $x^m N = (xN)^m = N$, so $x^m \in N$.

59. Suppose that $\mathrm{Aut}(G)$ is cyclic. Then $\mathrm{Inn}(G)$ is also cyclic. So, by Theorem 9.4, G/Z is cyclic and from Theorem 9.3 it follows that G is Abelian. This is a contradiction.

61. Say $|gH| = n$. Then $|g| = nt$ (by Exercise 37) and $|g^t| = n$. For the second part observe that every nonidentity element of Z has infinite order while $1 + \langle 3 \rangle$ has order 3 in $Z/\langle 3 \rangle$.

63. Note DK and VK are two of the four cosets but their product is not one of the four. So, closure fails. This does not contradict Theorem 9.2 because K is not normal.

65. First note that $|G/Z(G)| = |G|/|Z(G)| = 30/5 = 6$. By Theorem 7.2, the only groups of order 6 up to isomorphism are Z_6 and D_3. But $G/Z(G)$ can't be cyclic for if so, then by Theorem 9.3, G would be Abelian. In this case we would have $Z(G) = G$.

67. If A_5 had a normal subgroup of order 2 then, by Exercise 66, the subgroup has a nonidentity element that commutes with every

element of A_5. An element of A_5 of order 2 has the form $(ab)(cd)$. But $(ab)(cd)$ does not commute with (abc), which also belong to A_5.

69. Observe that $xg^2x^{-1} = (xgx^{-1})^2$.

71. If H is a subgroup of $A_4 \oplus Z_3$ of order 18, then the factor group $(A_4 \oplus Z_3)/H$ exists and has order 2. Then, for every element $(\beta, k) \in A_4 \oplus Z_3$ we have $((\beta, k)H)^2 = (\beta^2, 2k)H = H$. Thus, H contains every element of $A_4 \oplus Z_3$ of the form $(\beta^2, 2k)$, where $\beta \in A_4$ and $k \in Z_3$. However, in A_4 there are nine distinct elements of the form β^2 and in Z_3 there are three. This means that H has at least 27 elements.

CHAPTER 10
Group Homomorphisms

1. Note that $\det (AB) = (\det A)(\det B)$.

3. Note that $(f + g)' = f' + g'$.

5. Observe for every positive integer r we have $(xy)^r = x^r y^r$ so the mapping is a homomorphism. When r is odd the mapping is one-to-one and therefore an isomorphism. When n is even the mapping is two-to-one.

7. $(\sigma\phi)(g_1 g_2) = \sigma(\phi(g_1 g_2)) = \sigma(\phi(g_1)\phi(g_2)) = \sigma(\phi(g_1))\sigma(\phi(g_2)) = (\sigma\phi)(g_1)(\sigma\phi)(g_2)$.

9. $\phi((g,h)(g',h')) = \phi((gg',hh')) = gg' = \phi((g,h))\phi((g',h'))$. The kernel is $\{(e,h) \mid h \in H\}$.

11. The mapping $\phi : Z \oplus Z \to Z_a \oplus Z_b$ given by $\phi((x,y)) = (x \bmod a, y \bmod b)$ is operation preserving by Exercise 11 in Chapter 0. If $(x,y) \in \text{Ker } \phi$, then $x \in \langle a \rangle$ and $y \in \langle b \rangle$. So, $(x,y) \in \langle (a,0) \rangle \times \langle (0,b) \rangle$. Conversely, every element in $\langle (a,0) \rangle \times \langle (0,b) \rangle$ is in Ker ϕ. So, by Theorem 10.3, $Z \oplus Z \to Z_a \oplus Z_b$ is isomorphic to $\langle (a,0) \rangle \times \langle (0,b) \rangle$.

13. $(a,b) \to b$ is a homomorphism from $A \oplus B$ onto B with kernel $A \oplus \{e\}$. So, by Theorem 10.3, $(A \oplus B)/(A \oplus \{e\}) \approx B$.

15. By property 6 of Theorem 10.1, we know $\phi^{-1}(9) = 23 + \text{Ker } \phi = \{23, 3, 13\}$.

17. Suppose ϕ is such a homomorphism. By Theorem 10.3, Ker $\phi = \langle (8,1) \rangle, \langle (0,1) \rangle$ or $\langle (0,1) \rangle$. In these cases, the element $(1,0) + \text{Ker } \phi$ in $(Z_{16} \oplus Z_2)/\text{Ker } \phi$ has order either 16 or 8. So, $(Z_{16} \oplus Z_2)/\text{Ker } \phi$ is not isomorphic to $Z_4 \oplus Z_4$.

19. Since $| \text{Ker } \phi|$ is not 1 and divides 17, ϕ is the trivial map.

21. By Theorem 10.3 we know that $|Z_{30}/\text{Ker }\phi| = 5$. So, $|\text{Ker }\phi| = 6$. The only subgroup of Z_{30} of order 6 is $\langle 5 \rangle$.

23. **a.** The possible images are isomorphic to Z_1, Z_2, Z_3, Z_4, Z_6, and Z_{12}.

  b. $\langle 1 \rangle \approx Z_{36}$, $\langle 2 \rangle \approx Z_{18}$, $\langle 3 \rangle \approx Z_{12}$, $\langle 4 \rangle \approx Z_9$, $\langle 6 \rangle \approx Z_6$, and $\langle 12 \rangle \approx Z_3$.

25. To define a homomorphism from Z_{20} onto Z_{10} we must map 1 to a generator of Z_{10}. Since there are four generators of Z_{10} we have four homomorphisms. (Once we specific that 1 maps to an element a, the homomorphism is $x \rightarrow xa$.) To define a homomorphism from Z_{20} to Z_{10} we can map 1 to any element of Z_{10}. (Be careful here, these mappings are well defined only because 10 divides 20.)

27. For each k with $0 \le k \le n - 1$, the mapping $\phi(x) = kx$ is a homomorphism.

29. Say the kernel of the homomorphism is K. By Theorem 10.3, $G/K \approx Z_{10}$. So, $|G| = 10|K|$. In Z_{10}, let $\overline{H} = \langle 2 \rangle$. By properties 5, 7, and 8 of Theorem 10.2, $\phi^{-1}(\overline{H})$ is a normal subgroup of G of order $2|K|$. So, $\phi^{-1}(\overline{H})$ has index 2. To show that there is a subgroup of G of index 5, use the same argument with $\overline{H} = \langle 5 \rangle$.

31. By property 6 of Theorem 10.1, $\phi^{-1}(7) = 7\text{Ker }\phi = \{7, 17\}$.

33. By property 6 of Theorem 10.1,
$\phi^{-1}(11) = 11\text{Ker }\phi = \{11, 19, 27, 3\}$.

35. $\phi((a,b) + (c,d)) = \phi((a+b, c+d)) = (a+c) - (b+d) = a - b + c - d = \phi((a,b)) + \phi((c,d))$. $\text{Ker }\phi = \{(a,a) \mid a \in Z\}$.
$\phi^{-1}(3) = \{(a+3, a) \mid a \in Z\}$.

37. $\phi(xy) = (xy)^6 = x^6 y^6 = \phi(x)\phi(y)$. $\text{Ker }\phi = \langle \cos 60° + i \sin 60° \rangle$.

39. Consider the mapping ϕ from K to KN/N given by $\phi(k) = kN$. Since $\phi(kk') = kk'N = kNk'N = \phi(k)\phi(k')$ and $kN \in KN/N$, ϕ is a homomorphism. Moreover, $\text{Ker }\phi = K \cap H$. So, by Theorem 10.3, $K/(K \cap N) \approx KN/N$.

41. For each divisor d of k there is a unique subgroup of Z_k of order d, and this subgroup is generated by $\phi(d)$ elements. A homomorphism from Z_n to a subgroup of Z_k must carry 1 to a generator of the

subgroup. Furthermore, since the order of the image of 1 must divide n, so we need consider only those divisors d of k that also divide n.

43. Let N be a normal subgroup of D_4. By Lagrange's Theorem the only possibilities for $|N|$ are 1, 2, 4, and 8. By Theorem 10.4, the homomorphic images of D_4 are the same as the factor groups D_4/N of D_4. When $|N| = 1$, we know $N = \{e\}$ and $D_4/N \approx D_4$. When $|N| = 2$, then $N = \{R_0, R_{180}\}$, since this is the only normal subgroup of D_4 of order 2, and $D_4/N \approx Z_2 \oplus Z_2$ because D_4/N is a group of order 4 with three elements of order 2. When $n = 4$, $|D_4/N| = 2$ so $D_4/N \approx Z_2$. When $|N| = 8$, we have $D_4/N \approx \{e\}$.

45. It is divisible by 10.

47. It is infinite. Z

49. Let γ be a natural homomorphism from G onto G/N. Let $\overline{H}$ be a subgroup of G/N and let $\gamma^{-1}(\overline{H}) = H$. Then H is a subgroup of G and $H/N = \gamma(H) = \gamma(\gamma^{-1}(\overline{H})) = \overline{H}$.

51. The mapping $g \to \phi_g$ is a homomorphism with kernel $Z(G)$.

53. Since $(f + g)(3) = f(3) + g(3)$, the mapping is a homomorphism. The kernel is the set of elements in $Z[x]$ whose graphs pass through the point $(3, 0)$.

55. By Exercise 50 of Chapter 9, $H \cap K$ is a normal subgroup of G. By Exercise 39, $|H|/|H \cap K| = |HK|/|K| = |G|/|K| = 2$. Thus, $|G/(H \cap K)| = (|G|/|H|)(|H|/|H \cap K|) = 2 \cdot 2 = 4$. Since $H/(H \cap K)$ and $K/(H \cap K)$ are distinct subgroups of $G/(H \cap K)$ of order 2, $G/(H \cap K)$ is not cyclic.

57. Let G be a group of order 77. By Lagrange's Theorem every nonidentity of G has order 7, 11, or 77. If G has an element of order 77, then G is cyclic. So, we may assume that all nonidentity elements of G have order 7 or 11. Not all nonidentity elements can have order 11 because, by the Corollary of Theorem 4.4, the number of such elements is a multiple of 10. Not all nonidentity elements of G can have order 7 because the number of such elements is a multiple of 6. So, G must have elements a and b such that $|a| = 11$ and $|b| = 7$. Let $H = \langle a \rangle$. Then H is the only

subgroup of G of order 11 for if K is another one then by
Exercise 7 of the Supplementary Exercises for Chapters 5-8
$|HK| = |H||K|/|H \cap K| = 11 \cdot 11/1 = 121$. But HK is a subset of
G and G only has 77 elements. Because for every x in G, xHx^{-1} is
also a subgroup of G of order 11 (see Exercise 1 of the
Supplementary Exercises for Chapters 5-8), we must have
$xHx^{-1} = H$. So, $N(H) = G$. Since H has prime order, H is cyclic
and therefore Abelian. This implies that $C(H)$ contains H. So, 11
divides $|C(H)|$ and $|C(H)|$ divides 77. This implies that $C(H) = G$
or $C(H) = H$. If $C(H) = G$, then $|ab| = 77$. If $C(H) = H$, then
$|N(H)/C(H)| = 7$. But by the "N/C" Theorem (Example 15)
$N(H)/C(H)$ is isomorphic to a subgroup of $\text{Aut}(H)$
$\approx \text{Aut}(Z_{11}) \approx U(11)$ (see Theorem 6.5). Since $U(11) = 10$, we have
a contradiction.

59. Suppose that H is a proper subgroup of G that is not properly
contained in a proper subgroup of G. Then G/H has no nontrivial,
proper subgroup. It follows from Exercise 22 of Chapter 7 that
G/H is isomorphic to Z_p for some prime p. But then for every
coset $g + H$ we have $p(g + H) = H$ so that $pg \in H$ for all $g \in G$.
But then $G = pG \subseteq H$. Both Q and $\mathbf{R}$ satisfy the hypothesis.

CHAPTER 11

Fundamental Theorem of Finite Abelian Groups

1. $n = 4$
   $Z_4, Z_2 \oplus Z_2$

3. $n = 36$
   $Z_9 \oplus Z_4, Z_3 \oplus Z_3 \oplus Z_4, Z_9 \oplus Z_2 \oplus Z_2, Z_3 \oplus Z_3 \oplus Z_2 \oplus Z_2$

5. The only Abelian groups of order 45 are Z_{45} and $Z_3 \oplus Z_3 \oplus Z_5$. In the first group, $|3| = 15$; in the second one $|(1, 1, 1)| = 15$. $Z_3 \oplus Z_3 \oplus Z_5$ does not have an element of order 9.

7. In order to have exactly four subgroups of order 3, the group must have exactly 8 elements of order 3. When counting elements of order 3 we may ignore the components of the direct product that represent the subgroup of order 4 since their contribution is only the identity. Thus, we examine Abelian groups of order 27 to see which have exactly 8 elements of order 3. By Theorem 4.4, Z_{27} has exactly 2 elements of order 3; $Z_9 \oplus Z_3$ has exactly 8 elements of order 3 since for $|(a, b)| = 3$ we can choose $|a| = 1$ or 3 and $|b| = 1$ or 3, but not both $|a|$ and $|b|$ of order 1; In $Z_3 \oplus Z_3 \oplus Z_3$ every element except the identity has order 3. So, the Abelian groups of order 108 that have exactly four subgroups of order 3 are $Z_9 \oplus Z_3 \oplus Z_4$ and $Z_9 \oplus Z_3 \oplus Z_2 \oplus Z_2$. The subgroups of $Z_9 \oplus Z_3 \oplus Z_4$ of order 3 are $\langle (3, 0, 0) \rangle, \langle (0, 1, 0) \rangle, \langle (3, 1, 0) \rangle$ and $\langle (3, 2, 0) \rangle$. The subgroups of $Z_9 \oplus Z_3 \oplus Z_2 \oplus Z_2$ of order 3 are $\langle (3, 0, 0, 0) \rangle, \langle (0, 1, 0, 0) \rangle, \langle (3, 1, 0, 0) \rangle$ and $\langle (3, 2, 0, 0) \rangle$.

9. Elements of order 2 are determined by the factors in the direct product that have order a power of 2. So, we need only look at $Z_8, Z_4 \oplus Z_2$ and $Z_2 \oplus Z_2 \oplus Z_2$. By Theorem 4.4, Z_8 has exactly one element of order 2; $Z_4 \oplus Z_2$ has exactly three elements of order 2; $Z_2 \oplus Z_2 \oplus Z_2$ has exactly 7 elements of order 2. So, $G \approx Z_4 \oplus Z_2 \oplus Z_3 \oplus Z_5$.

11. By the Fundamental Theorem, any finite Abelian group G is isomorphic to some direct product of cyclic groups of prime-power order. Now go across the direct product and, for each distinct prime you have, pick off the largest factor of that prime-power. Next, combine all of these into one factor (you can do this, since their orders are relatively prime). Let us call the order of this new factor n_1. Now repeat this process with the remaining original factors and call the order of the resulting factor n_2. Then n_2 divides n_1, since each prime-power divisor of n_2 is also a prime-power divisor of n_1. Continue in this fashion. Example: If

$$G \approx Z_{27} \oplus Z_3 \oplus Z_{125} \oplus Z_{25} \oplus Z_4 \oplus Z_2 \oplus Z_2,$$

then

$$G \approx Z_{27 \cdot 125 \cdot 4} \oplus Z_{3 \cdot 25 \cdot 2} \oplus Z_2.$$

Now note that 2 divides $3 \cdot 25 \cdot 2$ and $3 \cdot 25 \cdot 2$ divides $27 \cdot 125 \cdot 4$.

13. $Z_2 \oplus Z_2$

15. a. 1   b. 1   c. 1   d. 1   e. 1   f. There is a unique Abelian group of order n if and only if n is not divisible by the square of any prime.

17. The symmetry group is $\{R_0, R_{180}, H, V\}$. Since this group is Abelian and has no element of order 4, it is isomorphic to $Z_2 \oplus Z_2$.

19. Because the group is Abelian and has order 9, the only possibilities are Z_9 and $Z_3 \oplus Z_3$. Since Z_9 has exactly 2 elements of order 3 and 9, 16, and 22 have order 3, the group must be isomorphic to $Z_3 \oplus Z_3$.

21. By the Corollary of Theorem 8.2, n must be square-free (no prime factor of n occurs more than once).

23. Among the first 11 elements in the table, there are 9 elements of order 4. None of the other isomorphism classes has this many.

25. First observe that G is Abelian and has order 16. Now we check the orders of the elements. Since the group has 8 elements of order 4 and 7 of order 2 it is isomorphic to $Z_4 \oplus Z_2 \oplus Z_2$.

27. Since Z_9 has exactly 2 elements of order 3 once we choose 3 nonidentity elements we will either have at least one element of

order 9 or 3 elements of order 3. In either case we have determined the group. The Abelian groups of order 18 are $Z_9 \oplus Z_2 \approx Z_{18}$ and $Z_3 \oplus Z_3 \oplus Z_2$. By Theorem 4.4, Z_{18} group has 6 elements of order 18, 6 elements of order 9, 2 of order 6, 2 of order 3, 1 of order 2, and 1 of order 1. $Z_3 \oplus Z_3 \oplus Z_2$ has 8 elements of order 3, 8 of order 6, 1 of order 2, and 1 of order 1. The worst case scenario is that at the end of 5 choices we have selected 2 of order 6, 2 of order 3, and 1 of order 2. In this case we still have not determined which group we have. But the sixth element we select will give us either an element of order 18 or 9, in which case we know the group Z_{18} or a third element of order 6 or 3, in which case we know the group is $Z_3 \oplus Z_3 \oplus Z_2$.

29. If $a^2 \neq b^2$, then $a \neq b$ and $a \neq b^3$. It follows that $\langle a \rangle \cap \langle b \rangle = \{e\}$. Then $G = \langle a \rangle \times \langle b \rangle \approx Z_4 \oplus Z_4$.

31. By Theorem 11.1, we can write the group in the form $Z_{p_1^{n_1}} \oplus Z_{p_2^{n_2}} \oplus \cdots \oplus Z_{p_k^{n_k}}$ where each p_i is an odd prime. By Theorem 8.1 the order of any element $(a_1, a_2, \ldots, a_k) = \operatorname{lcm}(|a_1|, |a_2|, \ldots, |a_k|)$. And from Theorem 4.3 we know that $|a_i|$ divides $p_i^{n_i}$, which is odd.

33. By Exercise 7 of the Supplementary Exercises for Chapters 5–8 we have, $|\langle a \rangle K| = |a||K|/|\langle a \rangle \cap K| = |a||K| = |\overline{a}||\overline{K}|p = |\overline{G}|p = |G|$.

35. By the Fundamental Theorem of Finite Abelian Groups, it suffices to show that every group of the form $Z_{p_1^{n_1}} \oplus Z_{p_2^{n_2}} \oplus \cdots \oplus Z_{p_k^{n_k}}$ is a subgroup of a U-group. Consider first a group of the form $Z_{p_1^{n_1}} \oplus Z_{p_2^{n_2}}$ (p_1 and p_2 need not be distinct). By Dirichlet's Theorem, for some s and t there are distinct primes q and r such that $q = tp_1^{n_1} + 1$ and $r = sp_2^{n_2} + 1$. Then $U(qr) = U(q) \oplus U(r) \approx Z_{tp_1^{n_1}} \oplus Z_{sp_2^{n_2}}$, and this latter group contains a subgroup isomorphic to $Z_{p_1^{n_1}} \oplus Z_{p_2^{n_2}}$. The general case follows in the same way.

SUPPLEMENTARY EXERCISES FOR CHAPTERS 9-11

1. Say $aH = Hb$. Then $a = hb$ for some h in H. Then
   $Ha = Hhb = Hb = aH$.

3. Suppose $\text{diag}(G)$ is normal. Then
   $(e,a)(b,b)(e,a)^{-1} = (b, aba^{-1}) \in \text{diag}(G)$. Thus $b = aba^{-1}$. If G is
   Abelian $(g,h)(b,b)(g,h)^{-1} = (gbg^{-1}, hbh^{-1}) = (b,b)$. The index of
   $\text{diag}(G)$ is $|G|$.

5. Let $\alpha \in \text{Aut}(G)$ and $\phi_a \in \text{Inn}(G)$. Then
   $(\alpha\phi_a\alpha^{-1})(x) = (\alpha\phi_a)(\alpha^{-1}(x)) = \alpha(a\alpha^{-1}(x)a^{-1}) = \alpha(a)x\alpha(a^{-1}) = \alpha(a)x(\alpha(a))^{-1} = \phi_{\alpha(a)}(x)$.

7. $\mathbf{R}^*$ (See Example 2 of Chapter 10.)

9. **a.** First we determine the center of H. Suppose that

$$\begin{bmatrix} 1 & a & b \\ 0 & 1 & c \\ 0 & 0 & 1 \end{bmatrix}$$

is in the center of H. Then for all choices of a', b' and c' we
have

$$\begin{bmatrix} 1 & a & b \\ 0 & 1 & c \\ 0 & 0 & 1 \end{bmatrix}\begin{bmatrix} 1 & a' & b' \\ 0 & 1 & c' \\ 0 & 0 & 1 \end{bmatrix} = \begin{bmatrix} 1 & a'+a & b'+ac'+b \\ 0 & 1 & c'+c \\ 0 & 0 & 1 \end{bmatrix}$$

And,

$$\begin{bmatrix} 1 & a' & b' \\ 0 & 1 & c' \\ 0 & 0 & 1 \end{bmatrix}\begin{bmatrix} 1 & a & b \\ 0 & 1 & c \\ 0 & 0 & 1 \end{bmatrix} = \begin{bmatrix} 1 & a+a' & b+a'c+b' \\ 0 & 1 & c+c' \\ 0 & 0 & 1 \end{bmatrix}.$$

So, $b' + ac' + b = b + a'c + b'$ for all choices of a', b' and c'.
Thus, $a'c = ac'$ for all choices of a' and c'. Taking $a' = 0$ and
$c' = 1$ gives $a = 0$. Taking $a' = 1$ and $c' = 0$ gives $c = 0$.

Finally, we note that $\begin{bmatrix} 1 & 0 & b \\ 0 & 1 & 0 \\ 0 & 0 & 1 \end{bmatrix}$ does commute with every element of H so $Z(H)$ consists of matrices of this form. Thus,

$$Z(H) = \left\{ \begin{bmatrix} 1 & 0 & b \\ 0 & 1 & 0 \\ 0 & 0 & 1 \end{bmatrix} \middle| b \in \mathbf{Q} \right\}.$$

The mapping

$$\phi\left(\begin{bmatrix} 1 & 0 & b \\ 0 & 1 & 0 \\ 0 & 0 & 1 \end{bmatrix} \right) = b$$

is an isomorphism. ϕ is 1-1 and onto by observation. To see that ϕ is operation preserving note that

$$\phi\left(\begin{bmatrix} 1 & 0 & b \\ 0 & 1 & 0 \\ 0 & 0 & 1 \end{bmatrix} \begin{bmatrix} 1 & 0 & b' \\ 0 & 1 & 0 \\ 0 & 0 & 1 \end{bmatrix} \right) = \phi\left(\begin{bmatrix} 1 & 0 & bb' \\ 0 & 1 & 0 \\ 0 & 0 & 1 \end{bmatrix} \right) = bb' =$$

$$\phi\left(\begin{bmatrix} 1 & 0 & b \\ 0 & 1 & 0 \\ 0 & 0 & 1 \end{bmatrix} \right) \phi\left(\begin{bmatrix} 1 & 0 & b' \\ 0 & 1 & 0 \\ 0 & 0 & 1 \end{bmatrix} \right).$$

b. To prove part b we define the mapping

$$\phi\left(\begin{bmatrix} 1 & a & b \\ 0 & 1 & c \\ 0 & 0 & 1 \end{bmatrix} \right) = (a, c).$$

By observation, ϕ is onto $Q \oplus Q$. To see that ϕ is operation preserving note that

$$\phi\left(\begin{bmatrix} 1 & a & b \\ 0 & 1 & c \\ 0 & 0 & 1 \end{bmatrix} \begin{bmatrix} 1 & a' & b' \\ 0 & 1 & c' \\ 0 & 0 & 1 \end{bmatrix} \right) =$$

$$\phi\left(\begin{bmatrix} 1 & a + a' & b' + ac' + b \\ 0 & 1 & c + c' \\ 0 & 0 & 1 \end{bmatrix} \right) = (a+a', c'+c) = (a, c)(a', c') =$$

$$\phi\left(\begin{bmatrix} 1 & a & b \\ 0 & 1 & c \\ 0 & 0 & 1 \end{bmatrix}\right)\phi\left(\begin{bmatrix} 1 & a' & b' \\ 0 & 1 & c' \\ 0 & 0 & 1 \end{bmatrix}\right).$$

Obviously, the kernel of ϕ is $Z(H)$. So, by Theorem 10.3, $H/Z(H) \approx Q \oplus Q$.

c. The proofs are valid for $\mathbf{R}$ and Z_p.

11. Every element of Q/Z can be written in the form $a/b + Z$ where b is a positive integer. Thus $b(a/b + Z) = a + Z = Z$ so that $a/b + Z$ has order at most $|b|$.

13. For $h \in H$ and $x \in G$ note that $(xhx^{-1})^n = xh^nx^{-1}$. So, $|xhx^{-1}| = n$ if and only if $|h| = n$. For H to be a subgroup n must be prime. To verify this note that if n is not prime we may write $n = mp$ where p is a prime and $m > 1$. But then if $h \in H$, we have that $|h^m| = p$ so that h^m is a not in H. In D_4, $|D| = 2$ and $|V| = 2$ but $DV = R_{90}$ has order 4.

15. Observe that $hkh^{-1}k^{-1} = (hkh^{-1})k^{-1} \in K$ and $hkh^{-1}k^{-1} = h(kh^{-1}k^{-1}) \in H$. So, $hkh^{-1}k^{-1} = e$ and therefore $hk = kh$.

17. By Theorem 7.3, $\text{stab}_G(5)$ has index 2 and by Exercise 7 of Chapter 9, $\text{stab}_G(5)$ is normal.

19. Suppose that ϕ is a homomorphism from $Z_8 \oplus Z_2 \oplus Z_2$ onto $Z_4 \oplus Z_4$. It follows from Theorem 10.3 that $(Z_8 \oplus Z_2 \oplus Z_2)/\text{Ker } \phi \approx Z_4 \oplus Z_4$ and $|\text{Ker } \phi| = 2$. Now observe that $\phi((4,0,0)) = \phi(4(1,0,0)) = 4\phi(1,0,0) = (0,0)$, so that $\text{Ker } \phi = \{(0,0,0),(4,0,0)\}$. However, in $(Z_8 \oplus Z_2 \oplus Z_2)/\text{Ker } \phi$ each of the four distinct elements $(1,0,0) + \text{Ker } \phi, (1,1,0) + \text{Ker } \phi, (1,0,1) + \text{Ker } \phi$, and $(1,1,1) + \text{Ker } \phi$ has order 4 whereas $Z_4 \oplus Z_4$ has only three elements of order 4.

21. Since $|S_4/H| = 6$, by Theorem 7.2, S_4/H is either cyclic or isomorphic to D_3 (which is isomorphic to S_3). But S_4 has no element of order 6 (see Exercise 7 in Chapter 5).

23. Recall that when n is even, $Z(D_n) = \{R_0, R_{180}\}$. For each of the n reflections F in D_n the element $FZ(D_n)$ in $D_n/Z(D_n)$ has order 2. The subgroup $\langle R_{360/n}\rangle/Z(D_n)$ has order $n/2 = m$. Since m is odd the only elements in D_m of order 2 are the m reflections.

25. The mapping $g \rightarrow g^n$ is a homomorphism from G onto G^n with kernel G_n. So, by Theorem 10.3, $G/G_n \approx G^n$.

27. Let $|H| = p$. Exercise 7 of Supplementary Exercises for Chapters 5–8 shows that H is the only subgroup of order p. But xHx^{-1} is also a subgroup of order p. So $xHx^{-1} = H$.

29. Say a and b are integers and $a/b + Z$ has order n in Q/Z. Then $na/b = m$ for some integer m. Thus,
$a/b + Z = m/n + Z = m(1/n + Z) \in \langle 1/n + Z \rangle$.

31. Let ϕ be a homomorphism from $Z \oplus Z$ into Z and let $\phi((1,0)) = a$ and let $\phi((0,1)) = b$. Then
$\phi((x,y)) = \phi(x(1,0) + y(0,1)) = x\phi((1,0)) + y\phi((0,1)) = ax + by$.

33. First note that by Exercise 11 of the Supplementary Exercises for Chapters 9–11, every element in Q/Z has finite order. For each positive integer n, let B_n denote the set of elements of order n and suppose that ϕ is an isomorphism from Q/Z to itself. Then, by property 5 of Theorem 6.2, $\phi(B_n) \subseteq B_n$. By Exercise 29 we know that B_n is finite, and since ϕ preserves orders and is one-to-one, we must have $\phi(B_n) = B_n$. Since it follows from Exercise 11 and Exercise 29 of Supplementary Exercises for Chapters 9–11 that $Q/Z = \bigcup B_n$, where the union is taken over all positive integers n, we have $\phi(Q/Z) = Q/Z$.

35. If the group is not Abelian, for any element a not in the center, the inner automorphism induced by a is not the identity. If the group is Abelian and contains an element a with $|a| > 2$, then $x \rightarrow x^{-1}$ works; if every nonidentity element has order 2, then G is isomorphic to a group of the form $Z_2 \oplus Z_2 \oplus \cdots \oplus Z_2$. In this case, the mapping that takes $(a_1, a_2, a_3, \ldots, a_k)$ to $(a_2, a_1, a_3, \ldots, a_k)$ is not the identity.

CHAPTER 12
Introduction to Rings

1. For any $n > 1$, the ring $M_2(Z_n)$ of 2×2 matrices with entries from Z_n is a finite noncommutative ring. The set $M_2(2Z)$ of 2×2 matrices with even integer entries is an infinite noncommutative ring that does not have a unity.

3. The only property that is not immediate is closure under multiplication:

$$(a + b\sqrt{2})(c + d\sqrt{2}) = (ac + 2bd) + (ad + bc)\sqrt{2}.$$

The proof is valid for every integer n.

5. The proofs given in Chapter 2 for a group apply to a ring as well.

7. First observe that every nonzero element a in Z_p has a multiplicative inverse a^{-1}. For part a, if $a \neq 0$, then $a^2 = a$ implies that $a^{-1}a^2 = a^{-1}a$ and therefore $a = 1$. For part b, if $a \neq 0$, then $ab = 0$ implies that $b = a^{-1}(ab) = a^{-1}0 = 0$. For part c, $ab = ac$ implies that $a^{-1}(ab) = a^{-1}(ac)$. So $b = c$.

9. If a and b belong to the intersection, then they belong to each member of the intersection. Thus $a - b$ and ab belong to each member of the intersection. So, $a - b$ and ab belong to the intersection.

11. Part 3:
$0 = 0(-b) = (a+(-a))(-b) = a(-b)+(-a)(-b) = -(ab)+(-a)(-b)$.
So, $ab = (-a)(-b)$.
Part 4:
$a(b - c) = a(b + (-c)) = ab + a(-c) = ab + (-(ac)) = ab - ac$.
Part 5: By Part 2, $(-1)a = 1(-a) = -a$.
Part 6: By Part 3, $(-1)(-1) = 1 \cdot 1 = 1$.

13. Let S be any subring of Z. By definition of a ring, S is a subgroup under addition. By Theorem 4.3, $S = \langle k \rangle$ for some integer k.

15. If m or n is 0 the statement follows from part 1 of Theorem 12.1. For simplicity, for any integer k and any ring element x we will use kx instead of $k \cdot x$. Then for positive m and n, observe that
$$(ma)(nb) = (a + a + \cdots + a)(b + b + \cdots + b) = (ab + ab + \cdots + ab),$$
where the terms $a = a + \cdots + a$, $b + b + \cdots + b$, and the last term have mn summands.

For the case that m is positive and n is negative, we first observe that nb means $(-b) + (-b) + \cdots + (-b) = (-n)(-b)$. So, $nb + (-n)b = [(-b) + (-b) + \cdots + (-b)][b + b + \cdots + b] = 0$. Thus, $0 = (ma)(nb + (-n)b) = (ma)(nb) + (ma)(-n)b = (ma)(nb) + m(-n)ab = (ma)(nb) + (-(mn))ab$. So, adding $(mn)ab$ to both ends of this string of equalities gives $(mn)ab = (ma)(nb)$. For the case when m is negative and n is positive just reverse the roles of m and n is the preceding argument. If both m and n are negative, note that
$$(ma)(nb) = ((-a) + (-a) + \cdots + (-a))((-b) + (-b) + \cdots + (-b)) = ((-m)(-a))((-n)(-b)) = (-m)(-n)((-a)(-b)) = (mn)(ab).$$

17. From Exercise 15, we have
$$(n \cdot a)(m \cdot a) = (nm) \cdot a^2 = (mn) \cdot a^2 = (m \cdot a)(n \cdot a).$$

19. Let a, b belong to the center. Then
$(a - b)x = ax - bx = xa - xb = x(a - b)$. Also,
$(ab)x = a(bx) = a(xb) = (ax)b = (xa)b = x(ab)$.

21. $(x_1, \ldots, x_n)(a_1, \ldots, a_n) = (x_1, \ldots, x_n)$ for all x_i in R_i if and only if $x_i a_i = x_i$ for all x_i in R_i and $i = 1, \ldots, n$ and $x_i a_i = x_i$ for all x_i in R_i if and only if x_i is a unity of R_i.

23. By observation $\pm 1, \pm i$ are units. To see that there are no others note that $(a + bi)^{-1} = \frac{1}{a+bi} = \frac{1}{a+bi} \frac{a-bi}{a-bi} = \frac{a-bi}{a^2+b^2}$. But $\frac{a}{a^2+b^2}$ is an integer only when $a^2 + b^2 = 1$ and this holds only when $a = \pm 1$ and $b = 0$ or $a = 0$ and $b = \pm 1$.

25. Note that the only $f(x) \in Z[x]$ for which $1/f(x)$ is a polynomial with integer coefficients are $f(x) = 1$ and $f(x) = -1$.

27. If a is a unit, then $b = a(a^{-1}b)$.

29. Note that $(a + b)(a^{-1} - a^{-2}b) = 1 - a^{-1}b + ba^{-1} - a^{-2}b^2 = 1$.

31. Let $a = \begin{bmatrix} 0 & 1 \\ 0 & 0 \end{bmatrix}$ and $b = \begin{bmatrix} 1 & 0 \\ 0 & 0 \end{bmatrix}$.

33. Note that $2x = (2x)^3 = 8x^3 = 8x$.

35. For Z_6 use $n = 3$. For Z_{10} use $n = 5$. Say $m = p^2 t$ where p is a prime. Then $(pt)^n = 0$ in Z_m since m divides $(pt)^n$.

37. Every subgroup of Z_n is closed under multiplication.

39. Since $ara - asa = a(r - s)a$ and $(ara)(asa) = ara^2sa = arsa$, S is a subring. Also, $a1a = a^2 = 1$, so $1 \in S$.

41. Let $\begin{bmatrix} a & a - b \\ a - b & b \end{bmatrix}$ and $\begin{bmatrix} a' & a' - b' \\ a' - b' & b' \end{bmatrix} \in R$. Then

$$\begin{bmatrix} a & a - b \\ a - b & b \end{bmatrix} - \begin{bmatrix} a' & a' - b' \\ a' - b' & b' \end{bmatrix} =$$

$$\begin{bmatrix} a - a' & (a - a') - (b - b') \\ (a - a') - (b - b') & b - b' \end{bmatrix} \in R. \text{ Also,}$$

$$\begin{bmatrix} a & a - b \\ a - b & b \end{bmatrix} \begin{bmatrix} a' & a' - b' \\ a' - b' & b' \end{bmatrix} =$$

$$\begin{bmatrix} aa' + aa' - ab' - ba' + bb' & aa' - bb' \\ aa' - bb' & aa' - ab' - ba' + bb' + bb' \end{bmatrix} \in R.$$

43. S is not a subring because $(1, 0, 1)$ and $(0, 1, 1)$ belong to S but $(1, 0, 1)(0, 1, 1) = (0, 0, 1)$ does not belong to S.

45. Observe that $n \cdot 1 - m \cdot 1 = (n - m) \cdot 1$. Also,
$(n \cdot 1)(m \cdot 1) = (nm) \cdot ((1)(1)) = (nm) \cdot 1$.

47. $S = \{m/2^n \mid m \in Z, n \in Z^+\}$ contains $1/2$. Since
$m/2^n - m'/2^{n'} = (m2^{n'} - 2^n m')/2^{n+n'} \in S$ and
$(m/2^n)(m'/2^{n'}) = mm'/2^{n+n'} \in S$, the subring test is satisfied. If T is any subring that contains $1/2$ then by closure under multiplication it contains $1/2^n$ and by closure under addition (and subtraction) it contains $m/2^n$. So, T contains S.

49. $(a + b)(a - b) = a^2 + ba - ab - b^2 = a^2 - b^2$ if and only if $ba - ab = 0$.

51. $Z_2 \oplus Z_2$; $Z_2 \oplus Z_2 \oplus \cdots$ (infinitely many copies).

CHAPTER 13
Integral Domains

1. For Example 1, observe that Z is a commutative ring with unity 1 and has no zero divisors. For Example 2, note that $Z[i]$ is a commutative ring with unity 1 and no zero divisors since it is a subset of $\mathbf{C}$, which has no zero divisors. For Example 3, note that $Z[x]$ is a commutative ring with unity $h(x) = 1$ and if $f(x) = a_n x^n + \cdots + a_0$ and $g(x) = b_m x^m + \cdots + b_0$ with $a_n \neq 0$ and $b_m \neq 0$, then $f(x)g(x) = a_n b_m x^{n+m} + \cdots + a_0 b_0$ and $a_n b_m \neq 0$. For Example 4, elements of $Z[\sqrt{2}]$ commute since they are real numbers; 1 is the unity; $(a + b\sqrt{2}) - (c + d\sqrt{2}) = (a - c) + (b - d)\sqrt{2}$ and $(a + b\sqrt{2})(c + d\sqrt{2}) = (ac + 2bd) + (bc + ad)\sqrt{2}$ so $Z[\sqrt{2}]$ is a ring; $Z[\sqrt{2}]$ has no zero divisors because it is a subring of $\mathbf{R}$, which has no zero divisors. For Example 5, note that Z_p is closed under addition and multiplication and multiplication is commutative; 1 is the unity; In Z_p, $ab = 0$ implies that p divides ab. So, by Euclid's Lemma (see Chapter 0), we know that p divides a or p divides b. Thus, in Z_p, $a = 0$ or $b = 0$. For Example 6, if n is not prime, then $n = ab$ where $1 < a < n$ and $1 < b < n$. But then $a \neq 0$ and $b \neq 0$ while $ab = 0$. For Example 7, note that

$$\begin{bmatrix} 1 & 0 \\ 0 & 0 \end{bmatrix} \begin{bmatrix} 0 & 0 \\ 0 & 1 \end{bmatrix} = \begin{bmatrix} 0 & 0 \\ 0 & 0 \end{bmatrix}.$$

For Example 8, note that $(1,0)(0,1) = (0,0)$.

3. Let $ab = 0$ and $a \neq 0$. Then $ab = a \cdot 0$, so $b = 0$.

5. Let $k \in Z_n$. If $\gcd(k, n) = 1$, then k is a unit. If $\gcd(k, n) = d > 1$, write $k = sd$. Then $k(n/d) = sd(n/d) = sn = 0$.

7. Let $s \in R$, $s \neq 0$. Consider the set $S = \{sr | r \in R\}$. If $S = R$, then $sr = 1$ (the unity) for some r. If $S \neq R$, then there are distinct r_1 and r_2 such that $sr_1 = sr_2$. In this case, $s(r_1 - r_2) = 0$. To see what happens when the "finite" condition is dropped, note that in the ring of integers 2 is neither a zero-divisor nor a unit.

9. $(a_1 + b_1\sqrt{d}) - (a_2 + b_2\sqrt{d}) = (a_1 - a_2) + (b_1 - b_2)\sqrt{d}$;
   $(a_1 + b_1\sqrt{d})(a_2 + b_2\sqrt{d}) = (a_1a_2 + b_1b_2d) + (a_1b_2 + a_2b_1)\sqrt{d}$. Thus the set is a ring. Since $Z[\sqrt{d}]$ is a subring of the ring of complex numbers, it has no zero-divisors.

11. The ring of even integers does not have a unity.

13. $(1 - a)(1 + a + a^2 + \cdots + a^{n-1}) =$
    $1 + a + a^2 + \cdots + a^{n-1} - a - a^2 - \cdots - a^n = 1 - a^n = 1 - 0 = 1$.

15. Suppose $a \neq 0$ and $a^n = 0$, where we take n to be as small as possible. Then $a \cdot 0 = 0 = a^n = a \cdot a^{n-1}$, so by cancellation, $a^{n-1} = 0$. This contradicts the assumption that n was as small as possible.

17. If $a^2 = a$ and $b^2 = b$, then $(ab)^2 = a^2b^2 = ab$.

19. $(3 + 4i)^2 = 3 + 4i$.

21. $a^2 = a$ implies $a(a - 1) = 0$. So if a is a unit, $a - 1 = 0$ and $a = 1$.

23. Since F is commutative so is K. The assumptions about K satisfy the conditions for the One-Step Subgroup Test for addition and for multiplication (excluding the 0 element). So, K is a subgroup under addition and a subgroup under multiplication (excluding 0). Thus K is a subring in which every nonzero element in a unit.

25. Note that $ab = 1$ implies $aba = a$. Thus $0 = aba - a = a(ba - 1)$. So, $ba - 1 = 0$.

27. A subdomain of an integral domain D is a subset of D that is an integral domain under the operations of D. To show that P is a subdomain, note that $n \cdot 1 - m \cdot 1 = (n - m) \cdot 1$ and $(n \cdot 1)(m \cdot 1) = (mn) \cdot 1$ so that P is a subring of D. Moreover, $1 \in P$, P has no zero divisors since D has none, and P is commutative because D is. Also, since every subdomain contains 1 and is closed under addition and subtraction, every subdomain contains P. Finally, we note that $|P| = $ char D when char D is prime and $|P|$ is infinite when char D is 0.

29. By Theorem 13.3, the characteristic is $|1|$. By Lagrange's Theorem (Theorem 7.1), $|1|$ divides 2^n. By Theorem 13.4, the characteristic is prime. Thus, the characteristic is 2.

31. **a.** First note that $a^3 = b^3$ implies that $a^6 = b^6$. Then $a = b$ because we can cancel a^5 from both sides (since $a^5 = b^5$).

    b. Since m and n are relatively prime, by Theorem 0.2, there are integers s and t such that $1 = sn + tm$. Since one of s and t is negative we may assume that s is negative. Then
    $$a(a^n)^{-s} = a^{1-sn} = (a^m)^t = (b^m)^t = b^{1-sn} = b(b^n)^{-s} = b(a^n)^{-s}.$$
    Now cancel $(a^n)^{-s}$.

33. $(1 - a)^2 = 1 - 2a + a^2 = 1 - 2a + a = 1 - a$.

35. Observe that $(1 + i)^4 = -1$, so $|1 + i| = 8$ and therefore the group is isomorphic to Z_8.

37. Let $S = \{a_1, a_2, \ldots, a_n\}$ be the nonzero elements of the ring. Then $a_1 a_1, a_1 a_2, \ldots, a_1 a_n$ are distinct elements for if $a_1 a_i = a_1 a_j$ then $a_1(a_i - a_j) = 0$ and therefore $a_i = a_j$. If follows that $S = \{a_1 a_1, a_1 a_2, \ldots, a_1 a_n\}$. Thus, $a_1 = a_1 a_i$ for some i. Then a_i is the unity, for if a_k is any element of S, we have $a_1 a_k = a_1 a_i a_k$, so that $a_1(a_k - a_i a_k) = 0$. Thus, $a_k = a_i a_k$ for all k.

39. Suppose that x and y are nonzero and $|x| = n$ and $|y| = m$ with $n < m$. Then $0 = (nx)y = x(ny)$. Since $x \neq 0$, we have $ny = 0$. This is a contradiction to the fact that $|y| = m$.

41. **a.** By the Binomial Theorem, $(x + y)^p = x^p + px^{p-1} + \cdots + px + 1$, where the coefficient of every term between x^p and 1 is divisible by p. Thus, $(x + y)^p = x^p + y^p$.

    b. We prove that $(x + y)^{p^n} = (x^{p^n} + y^{p^n})$ for every positive integer n by induction. The $n = 1$ case is done in part a. Assume that $(x + y)^{p^k} = x^{p^k} + y^{p^k}$. Then
    $$(x + y)^{p^{k+1}} = ((x + y)^{p^k})^p = (x^{p^k} + y^{p^k})^p = x^{p^{k+1}} + y^{p^{k+1}}.$$

    c. Let $S = \{0, 3, 6, 9\}$ in Z_{12}. Then S is a ring of characteristic 4 and $(3 + 3)^4 = 6^4 = 0$. But $3^4 + 3^4 = 9 + 9 = 6$.

43. By Theorems 13.3 and 13.4, $|1|$ has prime order, say p. Then by Exercise 39 every nonzero element has order p. If the order of the field were divisible by a prime q other than p, Theorem 9.5 implies that the field also has an element of order q. Thus, the order of the field is p^n for some prime p and some positive integer n.

45. $n \begin{bmatrix} a & b \\ c & d \end{bmatrix} = \begin{bmatrix} 0 & 0 \\ 0 & 0 \end{bmatrix}$ for all members of $M_2(R)$ if and only if $na = 0$ for all a in R.

47. This follows directly from Exercise 46.

49.   **a.** 2

   b. 2, 3

   c. 2, 3, 6, 11

   d. 2, 3, 9, 10

51. By Theorem 13.3, char R is prime. From $20 \cdot 1 = 0$ and $12 \cdot 1 = 0$ and Corollary 2 of Theorem 4.1, we know that char R divides both 12 and 20. Since the only prime that divides both 20 and 12 is 2, the characteristic is 2.

53. Note that $K = \{a + b\sqrt{2} \mid a, b \in Q\}$ is a field that contains $\sqrt{2}$ (see Example 10) and if F is any subfield of the reals that contains $\sqrt{2}$ then F contains K.

55. By Exercise 41, $x, y \in K$ implies that $x - y \in K$. Also, if $x, y \in K$ and $y \neq 0$, then $(xy^{-1})^p = x^p(y^{-1})^p = x^p(y^p)^{-1} = xy^{-1}$. So, by Exercise 23, K is a subfield.

57. Let $a \in F$, where $a \neq 0$ and $a \neq 1$. Then $(1 + a)^3 = 1^3 + 3(1^2 a) + 3(1 a^2) + a^3 = 1 + a + a^2 + a^3$. If $(1 + a)^3 = 1^3 + a^3$, then $a + a^2 = 0$. But then $a(1 + a) = 0$ so that $a = 0$ or $a = -1 = 1$. This contradicts our choice of a.

59. $\phi(x) = \phi(x \cdot 1) = \phi(x) \cdot \phi(1)$ so $\phi(1) = 1$. Also, $1 = \phi(1) = \phi(xx^{-1}) = \phi(x)\phi(x^{-1})$. So, $\phi(x) = 1$.

CHAPTER 14
Ideals and Factor Rings

1. Let r_1a and r_2a belong to $\langle a \rangle$. Then $r_1a - r_2a = (r_1 - r_2)a \in \langle a \rangle$. If $r \in R$ and $r_1a \in \langle a \rangle$, then $r(r_1a) = (rr_1)a \in \langle a \rangle$.

3. Clearly, I is not empty. Now observe that
$$(r_1a_1 + \cdots + r_na_n) - (s_1a_1 + \cdots + s_na_n) =$$
$$(r_1 - s_1)a_1 + \cdots + (r_n - s_n)a_n \in I.$$ Also, if $r \in R$, then
$r(r_1a_1 + \cdots + r_na_n) = (rr_1)a_1 + \cdots + (rr_n)a_n \in I.$ That $I \subseteq J$
follows from closure under addition and multiplication by elements
from R.

5. Let $a + bi, c + di \in S$. Then $(a + bi) - (c + di) = a - c + (b - d)i$ and
$b - d$ is even. Also, $(a + bi)(c + di) = ac - bd + (ad + cb)i$ and
$ad + cb$ is even. Finally, $(1 + 2i)(1 + i) = -1 + 3i \notin S$.

7. Since $ar_1 - ar_2 = a(r_1 - r_2)$ and $(ar_1)r = a(r_1r)$, aR is an ideal.
$4R = \langle 8 \rangle$

9. If n is a prime and $ab \in Z$ then by Euclid's Lemma (Chapter 0) n
divides a or n divides b. Thus, $a \in nZ$ or $b \in nZ$. If n is not a
prime, say $n = st$ where $s < n$ and $t < n$, then st belongs to nZ but
s and t do not.

11. **a.** $a = 1$ **b.** $a = 3$ **c.** $a = \gcd(m, n)$

13. **a.** $a = 12$

  b. $a = 48$. To see this, note that every element of $\langle 6 \rangle \langle 8 \rangle$ has the
  form $6t_18k_1 + 6t_28k_2 + \cdots + 6t_n8k_n = 48s \in \langle 48 \rangle$. So,
  $\langle 6 \rangle \langle 8 \rangle \subseteq \langle 48 \rangle$. Also, since $48 \in \langle 6 \rangle \langle 8 \rangle$, we have $\langle 48 \rangle \subseteq \langle 6 \rangle \langle 8 \rangle$.

  c. $a = mn$

15. Let $r \in R$. Then $r = 1r \in A$.

17. Let $u \in I$ be a unit and let $r \in R$. Then $r = r(u^{-1}u) = (ru^{-1})u \in I$.

19. I is closed under subtraction since the even integers are closed under subtraction. Also, if b_1, b_2, b_3, and b_4 are even, then every entry of $\begin{bmatrix} a_1 & a_2 \\ a_3 & a_4 \end{bmatrix} \begin{bmatrix} b_1 & b_2 \\ b_3 & b_4 \end{bmatrix}$ is even.

21. Use the observation that every member of R can be written in the form $\begin{bmatrix} 2q_1 + r_1 & 2q_2 + r_2 \\ 2q_3 + r_3 & 2q_4 + r_4 \end{bmatrix}$. Then note that
$$\begin{bmatrix} 2q_1 + r_1 & 2q_2 + r_2 \\ 2q_3 + r_3 & 2q_4 + r_4 \end{bmatrix} + I = \begin{bmatrix} r_1 & r_2 \\ r_3 & r_4 \end{bmatrix} + I.$$

23. $(br_1 + a_1) - (br_2 + a_2) = b(r_1 - r_2) + (a_1 - a_2) \in B$; $r'(br + a) = b(r'r) + r'a \in B$.

25. Suppose that I is an ideal of F and $I \neq \{0\}$. Let a be a nonzero element of I. Then by Exercise 17, $I = F$.

27. Since every element of $\langle x \rangle$ has the form $xg(x)$, we have $\langle x \rangle \subseteq I$. If $f(x) \in I$, then $f(x) = a_n x^n + \cdots + a_1 x = x(a_n x^{n-1} + \cdots + a_1) \in \langle x \rangle$.

29. Suppose $f(x) + A \neq A$. Then $f(x) + A = f(0) + A$ and $f(0) \neq 0$. Thus,
$$(f(x) + A)^{-1} = \frac{1}{f(0)} + A.$$
This shows that R/A is a field. Now use Theorem 14.4.

31. Since $(3 + i)(3 - i) = 10$ we know $10 + \langle 3 + i \rangle = 0 + \langle 3 + i \rangle$. Also, $i + \langle 3 + i \rangle = -3 + \langle 3 + i \rangle = 7 + \langle 3 + i \rangle$. Thus, every element $a + bi + \langle 3 + i \rangle$ can be written in the form $k + \langle 3 + i \rangle$ where $k = 0, 1, \ldots, 9$. Finally, $Z[i]/\langle 3 + i \rangle = \{k + \langle 3 + i \rangle \mid k = 0, 1, \ldots, 9\}$ since $1 + \langle 3 + i \rangle$ has additive order 10.

33. Note that in $(Z \oplus Z)/I$, $(a, b) + I = (a, 0) + (0, b) + I = (0, b) + I$. So, $(Z \oplus Z)/I$ is isomorphic to Z (map $(0, b) + I$ to b). Since Z is an integral domain but not a field, we have by Theorems 14.3 and 14.4 that I is a prime ideal but not a maximal ideal.

35. Since every element in $\langle x, 2 \rangle$ has the form $f(x) = xg(x) + 2h(x)$, we have $f(0) = 2h(0)$, so that $f(x) \in I$. If $f(x) \in I$, then $f(x) = a_n x^n + \cdots + a_1 x + 2k = x(a_n x^{n-1} \cdots + a_1) + 2k \in \langle x, 2 \rangle$. By Theorems 14.3 and 14.4 to prove that I is prime and maximal it suffices to show that $Z[x]/I$ is a field. To this end note that every

element of $Z[x]/I$ can be written in the form
$a_n x^n + \cdots + a_1 x + 2k + I = 0 + I$ or
$a_n x^n + \cdots + a_1 x + (2k+1) + I = 1 + I$. So, $Z[x]/I \approx Z_2$.

37. $3x + 1 + I$

39. Every ideal is a subgroup. Every subgroup of a cyclic group is cyclic.

41. Say $b, c \in \text{Ann}(A)$. Then $(b-c)a = ba - ca = 0 - 0 = 0$. Also, $(rb)a = r(ba) = r \cdot 0 = 0$.

43. a. $\langle 3 \rangle$
   b. $\langle 3 \rangle$
   c. $\langle 3 \rangle$

45. Suppose $(x + N(\langle 0 \rangle))^n = 0 + N(\langle 0 \rangle)$. We must show that $x \in N(\langle 0 \rangle)$. We know that $x^n + N(\langle 0 \rangle) = 0 + N(\langle 0 \rangle)$, so that $x^n \in N(\langle 0 \rangle)$. Then, for some $m, (x^n)^m = 0$, and therefore $x \in N(\langle 0 \rangle)$.

47. Let $I = \langle x^2 + x + 1 \rangle$. Then $Z_2[x]/I = \{0 + I, 1 + I, x + I, x + 1 + I\}$. $1 + I$ is its own multiplicative inverse and $(x+I)(x+1+I) = x^2 + x + I = x^2 + x + 1 + 1 + I = 1 + I$. So, every nonzero element of $Z_2[x]/I$ has a multiplicative inverse.

49. $x + 2 + \langle x^2 + x + 1 \rangle$ is not zero, but its square is.

51. If f and $g \in A$, then $(f-g)(0) = f(0) - g(0)$ is even and $(f \cdot g)(0) = f(0) \cdot g(0)$ is even. $f(x) = \sqrt{2} \in R$ and $g(x) = 2 \in A$, but $f(x)g(x) \notin A$.

53. Any ideal of R/I has the form A/I where A is an ideal of R. So, if $A = \langle a \rangle$, then $A/I = \langle a + I \rangle / I$.

55. By Theorem 14.3, R/I is an integral domain. Since every element in R/I is an idempotent and Exercise 16 in Chapter 13 says that the only idempotents in an integral domain are 0 and 1 we have that $R/I = \{0 + I, 1 + I\}$.

57. $\langle x \rangle \subset \langle x, 2^n \rangle \subset \langle x, 2^{n-1} \rangle \subset \cdots \subset \langle x, 2 \rangle$

59. Taking $r = 1$ and $s = 0$ shows that $a \in I$. Taking $r = 0$ and $s = 1$ shows that $b \in I$. If J is any ideal that contains a and b, then it contains I because of the closure conditions.

SUPPLEMENTARY EXERCISES FOR CHAPTERS 12-14

1. In Z_{10} they are 0, 1, 5, 6. In Z_{20}, they are 0, 1, 5, 16. In Z_{30}, they are 0, 1, 6, 10, 15, 16, 21, 25.

3. Suppose that $a^n = 0$ for some positive integer n. Let 2^k be the smallest power of 2 greater than n. Then $a^{2^k} = (a^{2^k-n})a^n = 0$. Since $0 = a^{2^k} = (a^{2^{k-1}})^2$ we conclude that $a^{2^{k-1}} = 0$. By iteration (or induction) we obtain that $a = 0$.

5. Suppose $A \not\subseteq C$ and $B \not\subseteq C$. Pick $a \in A$ and $b \in B$ so that $a, b \notin C$. But $ab \in C$ and C is prime. This contradicts the definition of a prime ideal.

7. Suppose that (a, b) is a nonzero element of an ideal I in $\mathbf{R} \oplus \mathbf{R}$. If $a \neq 0$, then $(r, 0) = (ra^{-1}, 0)(a, b) \in I$. Thus, $\mathbf{R} \oplus \{0\} \subseteq I$. Similarly, if $b \neq 0$, then $\{0\} \oplus \mathbf{R} \subseteq I$. So, the ideals of $\mathbf{R} \oplus \mathbf{R}$ are $\{0\} \oplus \{0\}, \mathbf{R} \oplus \mathbf{R}, \mathbf{R} \oplus \{0\}, \{0\} \oplus \mathbf{R}$. The ideals of $F \oplus F$ are $\{0\} \oplus \{0\}, F \oplus F, F \oplus \{0\}, \{0\} \oplus F$.

9. Consider $(m, n) + A$. Write $m = pq + r$ where $0 \leq r < p$. Then $(m, n) + A = (r, 0) + (pq, n) + A = (r, 0) + A$. So, $(Z \oplus Z)/A = \{(r, 0) \mid r = 0, 1, \ldots, p-1\} \approx Z_p$, which is a field. So, by Theorem 14.4 A is a maximal ideal.

11. Suppose that $A \subseteq B \cup C$ but $A \not\subseteq B$ and $A \not\subseteq C$. Let $a \in A$ but $a \notin B$ and $a' \in A$ but $a' \notin C$. Since $a \in A \subseteq B \cup C$ and $a \notin B$ we have that $a \in C$. Similarly, $a' \in B$. Then since $a + a' \in A$ we know that $a + a' \in B$ or $a + a' \in C$. If $a + a' \in B$ we have that $a = (a + a') - a' \in B$, which contradicts the fact that $a \notin B$. A similar contradiction arises if $a + a' \in C$.

13. Let $A = \{s_1 a t_1 + \cdots + s_n a t_n\}$. We first prove that A is a ideal of R. Since $(s_1 a t_1 + \cdots + s_n a t_n) - (s_1' a t_1' + \cdots + s_n' a t_n') = s_1 a t_1 + \cdots + s_n a t_n + (-s_1')a t_1' + \cdots + (-s_n')a t_n'$, A is closed under subtraction. If $r \in R$ then $r(s_1 a t_1 + \cdots + s_n a t_n) = (rs_1)a t_1 + \cdots + (rs_n)a t_n) \in A$ and $(s_1 a t_1 + \cdots + s_n a t_n)r = (s_1 a(t_1 r) + \cdots + s_n a(t_n r)) \in A$, so A is an

ideal of R that contains $1a1 = a$ (here 1 is the unity of R) and therefore $\langle a \rangle \subseteq A$. On the other hand, we know that $\langle a \rangle \supseteq A$ since multiplying a on the left or right by elements of R gives an element in $\langle a \rangle$ and $\langle a \rangle$ is closed under addition, it contains all elements of the form $s_1 a t_1 + \cdots + s_n a t_n$.

15. Since A is an ideal, $ab \in A$. Since B is an ideal, $ab \in B$. So $ab \in A \cap B = \{0\}$.

17. 6

19. Since for every nonzero element a in R, aR is a nonzero ideal of R, we have $aR = R$. Then, by Exercise 4, R is a field.

21. $x^2 + 1 + \langle x^4 + x^2 \rangle$.

23. In Z_8, $2^2 = 4 = 6^2$ and $2^3 = 0 = 6^3$.

25. Say char $R = p$ (remember p must be prime). Then char $R/A =$ the additive order of $1 + A$. But $|1 + A|$ divides $|1| = p$.

27. Let A be a prime ideal of R. By Theorem 14.3, R is an integral domain. Then, by Theorem 13.2, R/A is a field and, by Theorem 14.4, A is maximal.

29. Observe that $A = \left\{ \begin{bmatrix} a & b \\ 0 & 0 \end{bmatrix} \middle| \; a, b \in Z_2 \right\}$ but $\begin{bmatrix} 1 & 1 \\ 1 & 1 \end{bmatrix} \begin{bmatrix} 1 & 0 \\ 0 & 0 \end{bmatrix} = \begin{bmatrix} 1 & 0 \\ 1 & 0 \end{bmatrix}$ is not in A.

31. We claim that $R/A = \{0 + A, 1 + A\}$. To verify this claim let $a + bi + A$ be any element in R/A. If a and b are both even or both odd, then $a + bi + A = 0 + A$. If a is odd and b is even, then $a + bi + A = 1 + 0i + A = 1 + A$. If a is even and b is odd, then $a + bi + A = 0 + i + A = 1 + A$. So, $R/A = \{0 + A, 1 + A\}$, which is a field isomorphic to Z_2. Thus, by Theorem 14.4, A is maximal.

33. A finite subset of a field is a subfield if it contains a nonzero element and is closed under addition and multiplication.

35. Observe that $(a + bi)(a - bi) = a^2 + b^2 = 0$, so $a + bi$ is a zero-divisor. In $Z_{13}[i]$, $2 + 3i$ is a zero-divisor.

37. According to Theorem 13.3 we need only determine the additive order of $1 + \langle 2 + i \rangle$. Since
$$5(1 + \langle 2 + i \rangle) = 5 + \langle 2 + i \rangle = (2 + i)(2 - i) + \langle 2 + i \rangle = 0 + \langle 2 + i \rangle,$$
we know that $1 + \langle 2 + i \rangle$ has order 5.

39. The inverse is $2x + 3$.

41. Because elements of $Z_5[x, y]/\langle x, y \rangle$ have the form $f(x, y) + \langle x, y \rangle$ and $\langle x, y \rangle$ absorbs all terms of $f(x, y)$ that have an x or a y we know that $Z_5[x, y]/\langle x, y \rangle =$
$\{0 + \langle x, y \rangle, 1 + \langle x, y \rangle, 2 + \langle x, y \rangle, 3 + \langle x, y \rangle, 4 + \langle x, y \rangle\} \approx Z_5$. Thus, by Theorem 14.4, $\langle x, y \rangle$ is maximal.

43. Say $(a, b)^n = (0, 0)$. Then $a^n = 0$ and $b^n = 0$. If $a^m = 0$ and $b^n = 0$, then Thus, $(a, b)^{mn} = ((a^m)^n, (b^n)^m) = (0, 0)$.

45. If $a^2 = a$, then $p^k | a(a - 1)$. Since a and $a - 1$ are relatively prime, $p^k | a$ or $p^k | (a - 1)$. So, $a = 0$ or $a = 1$.

47. First observe that in Z_3, $a^2 + b^2 = 1$ or 2 except for the case that $a = 0 = b$. So, for any nonzero element $a + b\sqrt{2}$ in $Z_3[\sqrt{2}]$, $a^2 + b^2 = 1$ or -1. Next note that if $a + b\sqrt{2} \neq 0$, then
$$(a + b\sqrt{2})^{-1} = \frac{1}{a + b\sqrt{2}} = \frac{1}{a + b\sqrt{2}}\frac{a - b\sqrt{2}}{a - b\sqrt{2}} = \frac{a - b\sqrt{2}}{a^2 - 2b^2} =$$
$\frac{a - b\sqrt{2}}{a^2 + b^2} = a - b\sqrt{2}$ or $-a + b\sqrt{2}$. So, $Z_3[\sqrt{2}]$ is a field. In $Z_7[\sqrt{2}]$, $(1 + 2\sqrt{2})(1 + 5\sqrt{2}) = 0$.

CHAPTER 15
Ring Homomorphisms

1. Property 1: $\phi(nr) = n\phi(r)$ holds because a ring is a group under addition. To prove that $\phi(r^n) = (\phi(r))^n$ we note that by induction $\phi(r^n) = \phi(r^{n-1}r) = \phi(r^{n-1})\phi(r) = \phi(r)^{n-1}\phi(r) = \phi(r)^n$.
   Property 2: If $\phi(a)$ and $\phi(b)$ belong to $\phi(A)$ then $\phi(a) - \phi(b) = \phi(a - b)$ and $\phi(a)\phi(b) = \phi(ab)$ belong to $\phi(A)$.
   Property 3: $\phi(A)$ is a subgroup because ϕ is a group homomorphism. Let $s \in S$ and $\phi(r) = s$. Then $s\phi(a) = \phi(r)\phi(a) = \phi(ra)$ and $\phi(a)s = \phi(a)\phi(r) = \phi(ar)$.
   Property 4: Let a and b belong to $\phi^{-1}(B)$ and r belong to R. Then $\phi(a)$ and $\phi(b)$ are in B. So, $\phi(a) - \phi(b) = \phi(a) + \phi(-b) = \phi(a - b) \in B$. Thus, $a - b \in \phi^{-1}(B)$. Also, $\phi(ra) = \phi(r)\phi(a) \in B$ and $\phi(ar) = \phi(a)\phi(r) \in B$. So, ra and $ar \in \phi^{-1}(B)$.
   Property 5: $\phi(a)\phi(b) = \phi(ab) = \phi(ba) = \phi(b)\phi(a)$.
   Property 6: Because ϕ is onto, every element of S has the form $\phi(a)$ for some a in R. Then $\phi(1)\phi(a) = \phi(1a) = \phi(a)$ and $\phi(a)\phi(1) = \phi(a1) = \phi(a)$.
   Property 7: If ϕ is an isomorphism, by property 1 of Theorem 10.1 and the fact that ϕ is one-to-one, we have Ker $\phi = \{0\}$. If Ker $\phi = \{0\}$, by property 5 of Theorem 10.2, ϕ is one-to-one.
   Property 8: That ϕ^{-1} is one-to-one and preserves addition comes from property 3 of Theorem 6.3. To see that ϕ^{-1} preserves multiplication note that $\phi^{-1}(ab) = \phi^{-1}(a)\phi^{-1}(b)$ if and only if $\phi(\phi^{-1}(ab)) = \phi(\phi^{-1}(a)\phi^{-1}(b)) = \phi(\phi^{-1}(a))\phi(\phi^{-1}(b))$. But this reduces to $ab = ab$.

3. We already know $R/\text{Ker } \phi \approx \phi(R)$ as groups. Let $\Phi(x + \text{Ker } \phi) = \phi(x)$. Note that $\Phi((r + \text{Ker } \phi)(s + \text{Ker } \phi)) = \Phi(rs + \text{Ker } \phi) = \phi(rs) = \phi(r)\phi(s) = \Phi(r + \text{Ker } \phi)\Phi(s + \text{Ker } \phi)$.

5. $\phi(2 + 4) = \phi(1) = 5$, whereas $\phi(2) + \phi(4) = 0 + 0 = 0$.

7. **a.** No. Suppose that ϕ is a ring isomorphism from $2Z$ onto $3Z$

and let $\phi(2) = a$. Then
$\phi(4) = \phi(2 + 2) = \phi(2) + \phi(2) = a + a = 2a$ and
$\phi(4) = \phi(2 \cdot 2) = \phi(2)\phi(2) = a \cdot a = a^2$. But $2a = a^2$ implies
that $a = 0$ or $a = 2$, both of which give contradictions.

b. No. The argument given in part a applies in this case as well.

9. If a and $b \, (b \neq 0)$ belong to every member of the collection, then so
do $a - b$ and ab^{-1}. Thus, by Exercise 23 of Chapter 13, the
intersection is a subfield.

11. By observation ϕ is one-to-one and onto. Since
$$\phi((a + bi) + (c + di)) = \phi((a + c) + (b + d)i) = \begin{bmatrix} a + c & b + d \\ -(b + d) & a + c \end{bmatrix} =$$
$$\begin{bmatrix} a & b \\ -b & a \end{bmatrix} + \begin{bmatrix} c & d \\ -d & c \end{bmatrix} = \phi(a + bi) + \phi(c + di) \text{ addition is}$$
preserved. Also, $\phi((a + bi)(c + di)) = \phi((ac - bd) + (ad + bc)i) =$
$$\begin{bmatrix} ac - bd & ad + bc \\ -(ad + bc) & ac - bd \end{bmatrix} = \begin{bmatrix} a & b \\ -b & a \end{bmatrix} \begin{bmatrix} c & d \\ -d & c \end{bmatrix} =$$
$\phi(a + bi)\phi(c + di)$ so multiplication is preserved.

13. Since $\phi \left(\begin{bmatrix} a & b \\ c & d \end{bmatrix} \begin{bmatrix} a' & b' \\ c' & d' \end{bmatrix} \right) = \phi \left(\begin{bmatrix} aa' + bc' & ab' + bd' \\ ca' + dc' & cb' + dd' \end{bmatrix} \right) =$
$aa' + bc' \neq aa' = \phi \left(\begin{bmatrix} a & b \\ c & d \end{bmatrix} \right) \phi \left(\begin{bmatrix} a' & b' \\ c' & d' \end{bmatrix} \right)$ multiplication is
not preserved.

15. Yes. $\phi(x) = 6x$ is well defined because $a = b$ in Z_5 implies that 5
divides $a - b$. So, 30 divides $6a - 6b$. Moreover,
$\phi(a + b) = 6(a + b) = 6a + 6b = \phi(a) + \phi(b)$ and
$\phi(ab) = 6ab = 6 \cdot 6ab = 6a6b = \phi(a)\phi(b)$.

17. The set of all polynomials passing through the point $(1, 0)$.

19. Since $|A/B| = 4$ and $|2 + \langle 8 \rangle| = 4$, A/B is isomorphic to Z_4.
Because the ring A/B has no unity under multiplication, it is not
ring-isomorphic to Z_4.

21. Suppose that ϕ is a ring homomorphism from Z to Z and $\phi(1) = a$.
Then $\phi(2) = \phi(1 + 1) = 2\phi(1) = 2a$ and
$\phi(4) = \phi(2 + 2) = 2\phi(2) = 4a$. Also,
$\phi(4) = \phi(2 \cdot 2) = \phi(2)\phi(2) = 2a \cdot 2a = 4a^2$. Thus, $4a^2 = 4a$ and it

follows that $a = 0$ or $a = 1$. So, ϕ is the zero map or the identity map.

23. Suppose that ϕ is a ring homomorphism from $Z \oplus Z$ to $Z \oplus Z$. Let $\phi((1,0)) = (a,b)$ and $\phi((0,1)) = (c,d)$. Then
$\phi((x,y)) = \phi(x(1,0) + y(0,1)) = \phi(x(1,0)) + \phi(y(1,0)) = x\phi((1,0)) + y\phi((0,1)) = x(a,b) + y(c,d) = (ax + cy, bx + dy)$. Since ϕ preserves multiplication we know
$\phi((x,y)(x',y')) = \phi((xx',yy')) = (axx' + cyy', bxx' + dyy') = \phi((x,y))\phi(x',y')) = (ax + cy, bx + dy)(ax' + cy', bx' + dy') = ((ax + cy)(ax' + cy'), (bx + dy)(bx' + dy'))$. Now observe that $(ax + cy)(ax' + cy') = axx' + cyy'$ for all x, x', y, y' if and only if $a^2xx' + acxy' + acyx' + c^2yy' = axx' + cyy'$ for all x, x', y, y'. This implies that $a = 0$ or 1 and $c = 0$ or 1 and $ac = 0$. This gives $(a, c) = (0,0), (1,0)$, and $(0,1)$. Likewise, we have $(b, d) = (0,0), (1,0)$, and $(0,1)$. Thus, we have nine cases for (a,b,c,d):
$(0,0,0,0)$ corresponds to $(x,y) \rightarrow (0,0)$;
$(0,1,0,0)$ corresponds to $(x,y) \rightarrow (0,x)$;
$(0,0,0,1)$ corresponds to $(x,y) \rightarrow (0,y)$;
$(1,0,0,0)$ corresponds to $(x,y) \rightarrow (x,0)$;
$(1,1,0,0)$ corresponds to $(x,y) \rightarrow (x,x)$;
$(1,0,0,1)$ corresponds to $(x,y) \rightarrow (x,y)$;
$(0,0,1,0)$ corresponds to $(x,y) \rightarrow (y,0)$;
$(0,1,1,0)$ corresponds to $(x,y) \rightarrow (y,x)$;
$(0,0,1,1)$ corresponds to $(x,y) \rightarrow (y,y)$.
It is straight forward to show that each of these nine is a ring homomorphism.

25. Say 1 is the unity of R. Let $s = \phi(r)$ be any nonzero element of S. Then $\phi(1)s = \phi(1)\phi(r) = \phi(1r) = \phi(r) = s$. Similarly, $s\phi(1) = s$.

27. Suppose that ϕ is a ring homomorphism from $Z \oplus Z$ to Z. Let $\phi((1,0)) = a$ and $\phi((0,1)) = b$. Then
$\phi((x,y)) = \phi(x(1,0) + y(0,1)) = \phi(x(1,0)) + \phi(y(1,0)) = x\phi((1,0)) + y\phi((0,1)) = ax + by$. Since ϕ preserves multiplication we know
$\phi((x,y)(x',y')) = \phi((xx',yy')) = axx' + byy' = \phi((x,y)\phi(x',y')) = (ax + by)(ax' + by') = a^2xx' + abxy' + abyx' + b^2yy'$. Now observe that $axx' + byy' = a^2xx' + abxy' + abyx' + b^2yy'$ for all $x, x'y, y'$ if and only if $a^2 = a, b^2 = b$, and $ab = 0$. This means that $a = 0$ or 1

and $b = 0$ or 1 but not both $a = 1$ and $b = 1$. This gives us three cases for (a, b):

(0,0) corresponds to $(x, y) \to 0$;

(1,0) corresponds to $(x, y) \to x$;

(0,1) corresponds to $(x, y) \to y$.

Each of these is obviously a ring homomorphism.

29. Say $m = a_k a_{k-1} \cdots a_1 a_0$ and $n = b_k b_{k-1} \cdots b_1 b_0$. Then
$m - n = (a_k - b_k)10^k + (a_{k-1} - b_{k-1})10^{k-1} + \cdots + (a_1 - b_1)10 + (a_0 - b_0)$.
By the test for divisibility by 9 given in Example 8, $m - n$ is divisible by 9 provided that
$a_k - b_k + a_{k-1} - b_{k-1} + \cdots + a_1 - b_1 + a_0 - b_0 =$
$(a_k + a_{k-1} + \cdots + a_1 + a_0) - (b_k + b_{k-1} + \cdots + b_1 + b_0)$ is divisible by 9. But this difference is 0 since the second expression has the same terms as the first expression in some other order.

31. Since the sum of the digits of the number is divisible by 9 so is the number (see Example 8); the test for divisibility by 11 given in Exercise 30 is not satisfied.

33. Let α be the homomorphism from Z to Z_3 given by
$\alpha(n) = n \bmod 3$. Then noting that $\alpha(10^i) = \alpha(10)^i = 1^i = 1$ we have that $n = a_k a_{k-1} \cdots a_1 a_0 = a_k 10^k + a_{k-1} 10^{k-1} + \cdots + a_1 10 + a_0$ is divisible by 3 if and only if, modulo 3, $0 = \alpha(n) =$
$\alpha(a_k) + \alpha(a_{k-1}) + \cdots + \alpha(a_1) + \alpha(a_0) = \alpha(a_k + a_{k-1} + \cdots + a_1 + a_0)$.
But $\alpha(a_k + a_{k-1} + \cdots + a_1 + a_0) = 0 \bmod 3$ is equivalent $a_k + a_{k-1} + \cdots + a_1 + a_0$ being divisible by 3.

35. Observe that $10 \bmod 3 = 1$. So,
$(2 \cdot 10^{75} + 2) \bmod 3 = (2 + 2) \bmod 3 = 1$ and
$(10^{100} + 1) \bmod 3 = (1 + 1) \bmod 3 = 2 \bmod 3 = -1$. Thus,
$(2 \cdot 10^{75} + 2)^{100} \bmod 3 = 1^{100} \bmod 3 = 1$ and
$(10^{100} + 1)^{99} \bmod 3 = 2^{99} \bmod 3 = (-1)^{99} \bmod 3 = -1 = 2$.

37. By Theorem 13.3, the characteristic of R is the additive order of 1 and by property 6 of Theorem 15.1, the characteristic of S is the additive order of $\phi(1)$. Thus, by property 3 of Theorem 10.1, the characteristic of S divides the characteristic of R.

39. No. The kernel must be an ideal.

41. **a.** Suppose $ab \in \phi^{-1}(A)$. Then $\phi(ab) = \phi(a)\phi(b) \in A$, so that $a \in \phi^{-1}(A)$ or $b \in \phi^{-1}(A)$.

b. Let Φ be the homomorphism from R to S/A given by $\Phi(r) = \phi(r) + A$. Then $\phi^{-1}(A) = \text{Ker } \Phi$ and, by Theorem 15.3, $R/\text{Ker } \Phi \approx S/A$. So, $\phi^{-1}(A)$ is maximal.

43 **a.** Since $\phi((a,b) + (a',b')) = \phi((a+a', b+b')) = a + a' = \phi((a,b)) + \phi((a',b'))$, ϕ preserves addition. Also, $\phi((a,b)(a',b')) = \phi(aa', bb')) = aa' = \phi((a,b))\phi(a',b'))$ so ϕ preserves multiplication.

   b. $\phi(a) = \phi(b)$ implies that $(a,0) = (b,0)$, which implies that $a = b$. $\phi(a+b) = (a+b, 0) = (a, 0) + (b, 0) = \phi(a) + \phi(b)$. Also, $\phi(ab) = (ab, 0) = (a, 0)(b, 0) = \phi(a)\phi(b)$.

   c. Define ϕ by $\phi(r,s) = (s,r)$. By Exercise 14 in Chapter 8, ϕ is one-to-one and preserves addition. Since $\phi((r,s)(r',s')) = \phi((rr', ss')) = (ss', rr') = (s,r)(s', r') = \phi((r,s))\phi((r', s'))$ multiplication is also preserved.

45. Observe that $x^4 = 1$ has two solutions in **R** but four in **C**.

47. By Exercise 40 every ring homomorphism from **R** to **R** is an automorphism of **R**. And by Exercise 46 the only automorphism of **R** is the identity.

49. If $a/b = a'/b'$ and $c/d = c'/d'$, then $ab' = ba'$ and $cd' = dc'$. So, $acb'd' = (ab')(cd') = (ba')(dc') = bda'c'$. Thus, $ac/bd = a'c'/b'd'$ and therefore $(a/b)(c/d) = (a'/b')(c'/d')$.

51. Let F be the field of quotients of $Z[i]$. By definition $F = \{(a+bi)/(c+di)|\ a,b,c,d \in Z\}$. Since F is a field that contains Z and i, we know that $Q[i] \subseteq F$. But for any $(a+bi)/(c+di)$ in F we have $\frac{a+bi}{c+di} = \frac{a+bi}{c+di}\frac{c-di}{c-di} = \frac{(ac+bd)+(bc-ad)i}{c^2+d^2} = \frac{ac+bd}{c^2+d^2} + \frac{(bc-ad)i}{c^2+d^2} \in Q[i]$.

53. The subfield of E is $\{ab^{-1}|\ a, b \in D, b \neq 0\}$. Define ϕ by $\phi(ab^{-1}) = a/b$. Then $\phi(ab^{-1} + cd^{-1}) = \phi((ad + bc)(bd)^{-1})) = (ad + bc)/bd = ad/bd + bc/bd = a/b + c/d = \phi(ab^{-1}) + \phi(cd^{-1})$. Also, $\phi((ab^{-1})(cd^{-1})) = \phi(acb^{-1}d^{-1}) = \phi((ac)(bd)^{-1}) = ac/bd = (a/b)(c/d) = \phi(ab^{-1})\phi(cd^{-1})$.

55. Reflexive and symmetric properties follow from the commutativity of D. For transitivity, assume $a/b \equiv c/d$ and $c/d \equiv e/f$. Then $adf = (bc)f = b(cf) = bde$, and cancellation yields $af = be$.

57. Let ϕ be the mapping from T to Q given by $\phi(ab^{-1}) = a/b$. Now see Exercise 53.

59. Let $a_n x^n + a_{n-1} x^{n-1} + \cdots + a_0 \in \mathbf{R}[x]$ and suppose that $f(a + bi) = 0$. Then $a_n(a + bi)^n + a_{n-1}(a + bi)^{n-1} + \cdots + a_0 = 0$. By Example 2, the mapping ϕ from $\mathbf{C}$ to itself given by $\phi(a + bi) = a - bi$ is a ring isomorphism. So, by property 1 of Theorem 10.1,
$$0 = \phi(0) = \phi(a_n(a + bi)^n + a_{n-1}(a + bi)^{n-1} + \cdots + a_0) =$$
$$\phi(a_n)\phi((a + bi))^n + \phi(a_{n-1})\phi((a + bi))^{n-1} + \cdots + \phi(a_0) =$$
$$a_n(a - bi)^n + a_{n-1}(a - bi)^{n-1} + \cdots + a_0 = f(a - bi).$$

61. Certainly the unity 1 is contained in every subfield. So, if a field has characteristic p, the subfield $\{0, 1, \ldots, p - 1\}$ is contained in every subfield. If a field has characteristic 0, then $\{(m \cdot 1)(n \cdot 1)^{-1} \mid m, n \in Z, n \neq 0\}$ is a subfield contained in every subfield. This subfield is isomorphic to Q [map $(m \cdot 1)(n \cdot 1)^{-1}$ to m/n].

63. Observe that $(x + y)/1 = (x/1) + (y/1)$ and $(xy)/1 = (x/1)(y/1)$.

CHAPTER 16
Polynomial Rings

1. $f + g = 3x^4 + 2x^3 + 2x + 2$
   $ \, f \cdot g = 2x^7 + 3x^6 + x^5 + 2x^4 + 3x^2 + 2x + 2$

3. The zeros are 1, 2, 4, 5

5. Write $f(x) = (x - a)q(x) + r(x)$. Since $\deg(x - a) = 1$, $\deg r(x) = 0$ or $r(x) = 0$. So $r(x)$ is a constant. Also, $f(a) = r(a)$.

7. If a is a zero of $f(x)$ we know by Corollary 1 of Theorem 16.2 that the remainder when $f(x)$ is divided by $x - a$ is 0. So, $x - a$ is a factor of $f(x)$. If $x - a$ is a factor of $f(x)$, then the remainder when $f(x)$ is divided by $x - a$ is 0. So, by Corollary 1 of Theorem 16.2, $f(a) = 0$.

9. Let $f(x)$, $g(x) \in R[x]$. By inserting terms with the coefficient 0 we may write

   $$f(x) = a_n x^n + \cdots + a_0 \quad \text{and} \quad g(x) = b_n x^n + \cdots + b_0.$$

   Then

   $$
   \begin{aligned}
   \overline{\phi}(f(x) + g(x)) &= \phi(a_n + b_n)x^n + \cdots + \phi(a_0 + b_0) \\
   &= (\phi(a_n) + \phi(b_n))x^n + \cdots + \phi(a_0) + \phi(b_0) \\
   &= (\phi(a_n)x^n + \cdots + \phi(a_0)) + (\phi(b_n)x^n + \cdots + \phi(b_0)) \\
   &= \overline{\phi}(f(x)) + \overline{\phi}(g(x)).
   \end{aligned}
   $$

   Multiplication is done similarly.

11. The quotient is $2x^2 + 2x + 1$; the remainder is 2.

13. Observe that $(2x + 1)(2x + 1) = 4x^2 + 4x + 1 = 1$. So, $2x + 1$ is its own inverse.

15. No, because if $a_n, b_m \in Z_p$ and are not zero, then
    $(a_n x^n + \cdots + a_0)(b_m x^m + \cdots + b_0) = a_n b_m x^{n+m} + \cdots + a_0 b_0$ and
    $a_n b_m \neq 0$.

17. If $f(x) = a_n x^n + \cdots + a_0$ and $g(x) = b_m x^m + \cdots + b_0$, then
    $f(x) \cdot g(x) = a_n b_m x^{m+n} + \cdots + a_0 b_0$ and $a_n b_m \neq 0$ when $a_n \neq 0$ and
    $b_m \neq 0$.

19. Suppose that $f(x) \neq 0$ and $f(x)$ is a polynomial of degree n. Then,
    by Corollary 3 of Theorem 16.2, $f(x)$ has at most n zeros. This is a
    contradiction to the assumption that $f(x)$ has infinitely many zeros.

21. If $f(x) \neq g(x)$, then $\deg[f(x) - g(x)] < \deg p(x)$. But the minimum
    degree of any member of $\langle p(x) \rangle$ is $\deg p(x)$. So, $f(x) - g(x)$ does
    not have a degree. This means that $f(x) - g(x) = 0$.

23. We start with $(x - 1/2)(x + 1/3)$ and clear fractions to obtain
    $(6x - 3)(6x + 2)$ as one possible solution.

25. The proof given for Theorem 16.2 with $g(x) = x - a$ is valid over
    any commutative ring with unity. Moreover, the proofs for
    Corollaries 1 and 2 of Theorem 16.2 are also valid over any
    commutative ring with unity.

27. Observe that $f(x) \in I$ if and only if $f(1) = 0$. Then if f and g
    belong to I and h belongs to $F[x]$, we have
    $(f - g)(1) = f(1) - g(1) = 0 - 0$ and
    $(hf)(1) = h(1)f(1) = h(1) \cdot 0 = 0$. So, I is an ideal. By
    Theorem 16.4, $I = \langle x - 1 \rangle$.

29. This follows directly from Corollary 2 of Theorem 15.5 and
    Exercise 9 in this chapter.

31. For any a in $U(p)$, $a^{p-1} = 1$, so every member of $U(p)$ is a zero of
    $x^{p-1} - 1$. From the Factor Theorem (Corollary 2 of Theorem 16.2)
    we obtain that $g(x) = (x - 1)(x - 2) \cdots (x - (p - 1))$ is a factor of
    $x^{p-1} - 1$. Since both $g(x)$ and $x^{p-1} - 1$ have lead coefficient 1, the
    same degree, and their difference has $p - 1$ zeros, their difference
    must be 0 (for otherwise their difference would be a polynomial of
    degree less than $p - 1$ that had $p - 1$ zeros).

33. By Exercise 32, $(p - 1)! \bmod p = p - 1$. So, p divides
    $(p - 1)! - (p - 1) = (p - 1)((p - 2)! - 1)$. Since p does not divide

$(p-1)$ we know that p divides $(p-2)! - 1$. Thus, $(p-2)! \bmod p = 1$.

35. Observe that, modulo 101,
$$(50!)^2 = (50!)(-1)(-2)\cdots(-50) = (50!)(100)(99)\cdots(51) = 100!.$$
And by Exercise 32, $100! \bmod 101 = 100 = -1 \bmod 101$.

37. Note that $I = \langle 2 \rangle$ is maximal in Z but $I[x]$ is not maximal in $Z[x]$ since $I[x]$ is properly contained in the ideal
$\{f(x) \in Z[x] \mid f(0) \text{ is even }\}$.

39. Since $F[x]$ is a PID, $\langle f(x), g(x) \rangle = \langle a(x) \rangle$ for some $a(x) \in F[x]$. Thus $a(x)$ divides both $f(x)$ and $g(x)$. This means that $a(x)$ is a constant. So, by Exercise 17 in Chapter 14, $\langle f(x), g(x) \rangle = F[x]$. Thus, $1 \in \langle f(x), g(x) \rangle$.

41. By the Factor Theorem (Corollary 2 of Theorem 16.2) we may write $f(x) = (x-a)g(x)$. Then $f'(x) = (x-a)g'(x) + g(x)$. Thus, $g(a) = 0$ and by the Factor Theorem $x - a$ is a factor of $g(x)$.

43. Say $\deg g(x) = m, \deg h(x) = n$, and $g(x)$ has leading coefficient a. Let $k(x) = g(x) - ax^{m-n}h(x)$. Then $\deg k(x) < \deg g(x)$ and $h(x)$ divides $k(x)$ in $Z[x]$ by induction. So, $h(x)$ divides $k(x) + ax^{m-n}h(x) = g(x)$ in $Z[x]$.

45. By the Division Algorithm (Theorem 16.2) we may write $x^{43} = (x^2 + x + 1)q(x) + r(x)$ where $r(x) = 0$ or $\deg r(x) < 2$. Thus, $r(x)$ has the form $cx + d$. Then $x^{43} - cx - d$ is divisible by $x^2 + x + 1$. Finally, let $a = -c$ and $b = -d$.

47. Observe that every term of $f(a)$ has the form $c_i a^i$ and $c_i a^i \bmod m = c_i b^i \bmod m$. To prove the second statement assume that there is some integer k so that $f(k) = 0$. If k is even, then because $k \bmod 2 = 0 \bmod 2$, we have by the first statement $0 = f(k) \bmod 2 = f(0) \bmod 2$ so that $f(0)$ is even. This shows that k is not even. If k is odd, then $k \bmod 2 = 1 \bmod 2$, so by the first statement $f(k) = 0$ is odd. This contradiction completes the proof.

49. A solution to $x^{25} - 1 = 0$ in Z_{37} is a solution to $x^{25} = 1$ in $U(37)$. So, by Corollary 2 of Theorem 4.1, $|x|$ divides 25. Moreover, we must also have that $|x|$ divides $|U(37)| = 36$. So, $|x| = 1$ and therefore $x = 1$.

CHAPTER 17
Factorization of Polynomials

1. By Theorem 17.1, $f(x)$ is irreducible over **R**. Over **C** we have
$2x^2 + 4 = 2(x^2 + 2) = 2(x + \sqrt{2}i)(x - \sqrt{2}i)$.

3. If $f(x)$ is not primitive, then $f(x) = ag(x)$, where a is an integer greater than 1. Then a is not a unit in $Z[x]$ and $f(x)$ is reducible.

5.   **a.** If $f(x) = g(x)h(x)$, then $af(x) = ag(x)h(x)$.
    b. If $f(x) = g(x)h(x)$, then $f(ax) = g(ax)h(ax)$.
    c. If $f(x) = g(x)h(x)$, then $f(x + a) = g(x + a)h(x + a)$.
    d. Let $f(x) = 8x^3 - 6x + 1$. Then
$f(x + 1) = 8(x + 1) - 6(x + 1) + 1 =$
$8x^3 + 24x^2 + 24x + 8 - 6x - 6 + 1 = 8x^3 + 24x^2 - 18x + 3$. By Eisenstein's Criterion (Theorem 17.4), $f(x + 1)$ is irreducible over Q and by part c, $f(x)$ is irreducible over Q.

7. It follows from Theorem 17.1 that $p(x) = x^2 + x + 1$ is irreducible over Z_5. Then from Corollary 1 of Theorem 17.5 we know that $Z_5[x]/\langle p(x)\rangle$ is a field. To see that this field has order 25 note that if $f(x) + \langle p(x)\rangle$ is any element of $Z_5[x]/\langle p(x)\rangle$, then by the Division Algorithm (Theorem 16.2) we may write $f(x) + \langle p(x)\rangle$ in the form $p(x)q(x) + ax + b + \langle p(x)\rangle = ax + b + \langle p(x)\rangle$. Moreover, $ax + b + \langle p(x)\rangle = cx + d + \langle p(x)\rangle$ only if $a = c$ and $b = d$, since $(a - c)x + b - d$ is divisible by $\langle p(x)\rangle$ only when it is 0. So, $Z_5[x]/\langle p(x)\rangle$ has order 25.

9. Note that -1 is a zero. No, since 4 is not a prime.

11. Let $f(x) = x^4 + 1$ and $g(x) = f(x + 1) = x^4 + 4x^3 + 6x^2 + 4x + 2$. Then $f(x)$ is irreducible over Q if $g(x)$ is. Eisenstein's Criterion shows that $g(x)$ is irreducible over Q. To see that $x^4 + 1$ is reducible over **R**, observe that

$$x^8 - 1 = (x^4 + 1)(x^4 - 1)$$

so any complex zero of $x^4 + 1$ is a complex zero of $x^8 - 1$. Also note that the complex zeros of $x^4 + 1$ must have order 8 (when considered as an element of $\mathbf{C}$). Let $\omega = \sqrt{2}/2 + i\sqrt{2}/2$. Then Example 2 in Chapter 16 tells us that the complex zeros of $x^4 + 1$ are $\omega, \omega^3, \omega^5$, and ω^7, so

$$x^4 + 1 = (x - \omega)(x - \omega^3)(x - \omega^5)(x - \omega^7).$$

But we may pair these factors up as:

$$((x - \omega)(x - \omega^7))((x - \omega^3)(x - \omega^5))$$
$$= (x^2 - \sqrt{2}x + 1)(x^2 + \sqrt{2}x + 1)$$

to factor using reals (see DeMoivre's Theorem, Example 8 in Chapter 0).

13. $(x + 3)(x + 5)(x + 6)$

15. **a.** Since every reducible polynomial of the form $x^2 + ax + b$ can be written in the form $(x - c)(x - d)$ we need only count the number of distinct such expressions over Z_p. Note that there are $p(p - 1)$ expressions of the form $(x - c)(x - d)$ where $c \neq d$. However, since $(x - c)(x - d) = (x - d)(x - c)$ there are only $p(p - 1)/2$ distinct such expressions. To these we must add the p cases of the form $(x - c)(x - c)$. This gives us $p(p - 1)/2 + p = p(p + 1)/2$.

b. First note that for every reducible polynomial of the form $f(x) = x^2 + ax + b$ over Z_p the polynomial $cf(x)$ $(c \neq 0)$ is also reducible over Z_p. By part a, this gives us at least $(p - 1)p(p + 1)/2 = p(p^2 - 1)/2$ reducible polynomials over Z_p. Conversely, every quadratic polynomial over Z_p can be written in the form $cf(x)$ where $f(x)$ has lead coefficient 1. So, the $p(p^2 - 1)/2$ reducibles we have already counted includes all cases.

17. By Exercise 16, for each prime p there is an irreducible polynomial $p(x)$ of degree 2 over Z_p. By Corollary 1 of Theorem 17.5, $Z_p[x]/\langle p(x) \rangle$ is a field. By the Division Algorithm (Theorem 16.2) every element in $Z_p[x]/\langle p(x) \rangle$ can be written in the form $ax + b + \langle p(x) \rangle$. Moreover, $ax + b + \langle p(x) \rangle = cx + d + \langle (p(x) \rangle$ only when $a = c$ and $c = d$ since $(ax + b) - (cx + d)$ is divisible by $p(x)$ only when it is 0. Thus, $Z_p[x]/\langle p(x) \rangle$ has order p^2.

19. Consider the mapping from $Z_3[x]$ onto $Z_3[i]$ given by
$\phi(f(x)) = f(i)$. Since $\phi(f(x) + g(x)) = \phi((f + g)(x)) = (f + g)(i) = f(i) + g(i) = \phi(f(x)) + \phi(g(x))$ and
$\phi(f(x)g(x)) = \phi((fg)(x) = (fg)(i) = f(i)g(i) = \phi(f(x))\phi(g(x))$, ϕ
is a ring homomorphism. Because $\phi(x^2 + 1) = i^2 + 1 = -1 + 1 = 0$
we know that $x^2 + 1 \in \text{Ker } \phi$. From Theorem 16.4 we have that
Ker $\phi = \langle x^2 + 1 \rangle$. Finally, Theorem 15.3 gives us that
$Z_3[x]/\langle x^2 + 1 \rangle \approx Z_3[i]$.

21. $x^2 + 1, x^2 + x + 2, x^2 + 2x + 2$

23. 1 has multiplicity 1, 3 has multiplicity 2.

25. We know that $a_n(r/s)^n + a_{n-1}(r/s)^{n-1} + \cdots + a_0 = 0$. So, clearing
fractions we obtain $a_n r^n + s a_{n-1} r^{n-1} + \cdots + s^n a_0 = 0$. This shows
that $s \mid a_n r^n$ and $r \mid s^n a_0$. By Euclid's Lemma (Chapter 0), s
divides a_n or s divides r^n. Since s and r are relatively prime, s
must divide a_n. Similarly, r must divide a_0.

27. Since $a_1(x)a_2(x) \cdots a_k(x) = a_1(x)(a_2(x) \cdots a_k(x))$, we have by
Corollary 2 of Theorem 17.5 that $p(x)$ divides $a_1(x)$ or $p(x)$ divides
$a_2(x) \cdots a_k(x)$. In the latter case, the Second Principle of
Mathematical Induction implies that $p(x)$ divides some $a_i(x)$ for
$i = 2, 3, \ldots, k$.

29. If there is an a in Z_p such that $a^2 = -1$, then
$x^4 + 1 = (x^2 + a)(x^2 - a)$.

If there is an a in Z_p such that $a^2 = 2$, then
$x^4 + 1 = (x^2 + ax + 1)(x^2 - ax + 1)$.

If there is an a in Z_p such that $a^2 = -2$, then
$x^4 + 1 = (x^2 + ax - 1)(x^2 - ax - 1)$.

To show that one of these three cases must occur, consider the
homomorphism from Z_p^* to itself given by $x \rightarrow x^2$. Since the kernel
is $\{1, -1\}$, the image H has index 2 (we may assume that $p \neq 2$).
Suppose that neither -1 nor 2 belongs to H. Then, since there is
only 1 coset other than H, we have $-1H = 2H$. Also, since Z_p^*/H
has order 2, any element in Z_p^*/H squared is the identity coset H.
Thus, $H = (-1H)(-1H) = (-1H)(2H) = -2H$, so that -2 is in
H.

31. Since $(f+g)(a) = f(a) + g(a)$ and $(f \cdot g)(a) = f(a)g(a)$, the mapping is a homomorphism. Clearly, $p(x)$ belongs to the kernel. By Theorem 17.5, $\langle p(x) \rangle$ is a maximal ideal, so the kernel is $\langle p(x) \rangle$.

33. Consider the mapping ϕ from F to $F[x]/\langle p(x) \rangle$ given by $\phi(a) = a + \langle p(x) \rangle$. By observation, ϕ is one-to-one and onto. Moreover,
$\phi(a+b) = a + b + \langle p(x) \rangle = a + \langle p(x) \rangle + b + \langle p(x) \rangle = \phi(a) + \phi(b)$
and $\phi(ab) = ab + \langle p(x) \rangle = (a + \langle p(x) \rangle)(b + \langle p(x) \rangle) = \phi(a)\phi(b)$ so ϕ is a ring isomorphism.

35. The analysis is identical except that $0 \le q, r, t, u \le n$. Now just as when $n = 2$, we have $q = r = t = 1$, but this time $0 \le u \le n$. However, when $u > 2$, $P(x) = x(x+1)(x^2 + x + 1)(x^2 - x + 1)^u$ has $(-u + 2)x^{2u+3}$ as one of its terms. Since the coefficient of x^{2u+3} represents the number of dice with the label $2u + 3$, the coefficient cannot be negative. Thus, $u \le 2$, as before.

37. Although the probability of rolling any particular sum is the same with either pair of dice, the probability of rolling doubles is different (1/6 with ordinary dice, 1/9 with Sicherman dice). Thus, the probability of going to jail is different. Other probabilities are also affected. For example, if in jail one cannot land on Virginia by rolling a pair of 2s with Sicherman dice, but one is twice as likely to land on St. James with a pair of 3s with the Sicherman dice as with ordinary dice.

CHAPTER 18
Divisibility in Integral Domains

1.   1. $|a^2 - db^2| = 0$ implies $a^2 = db^2$. Thus $a = 0 = b$, since otherwise $d = 1$ or d is divisible by the square of a prime.

   2. $N((a + b\sqrt{d})(a' + b'\sqrt{d})) = N(aa' + dbb' + (ab' + a'b)d) = |(a^2 - db^2)(a'^2 - db'^2)| = |(aa' + dbb')^2 - d(ab' + a'b)^2| = |a^2a'^2 + d^2b^2b'^2 - da^2b'^2 - da'^2b^2| = |a^2 - db^2||a'^2 - db'^2| = N(a + b\sqrt{d})N(a' + b'\sqrt{d})$.

   3. If $xy = 1$, then $1 = N(1) = N(xy) = N(x)N(y)$ and $N(x) = 1 = N(y)$. If $N(a + b\sqrt{d}) = 1$, then $\pm 1 = a^2 - db^2 = (a + b\sqrt{d})(a - b\sqrt{d})$ and $a + b\sqrt{d}$ is a unit.

   4. This part follows directly from 2 and 3.

3. Let $I = \cup I_i$. Let $a, b \in I$ and $r \in R$. Then $a \in I_i$ for some i and $b \in I_j$ for some j. Thus $a, b \in I_k$, where $k = \max\{i, j\}$. So, $a - b \in I_k \subseteq I$ and ra and $ar \in I_k \subseteq I$.

5. Clearly, $\langle ab \rangle \subseteq \langle b \rangle$. If $\langle ab \rangle = \langle b \rangle$, then $b = rab$, so that $1 = ra$ and a is a unit.

7. Say $x = a + bi$ and $y = c + di$. Then

$$xy = (ac - bd) + (bc + ad)i.$$

So

$$d(xy) = (ac - bd)^2 + (bc + ad)^2 = (ac)^2 + (bd)^2 + (bc)^2 + (ad)^2.$$

On the other hand,

$$d(x)d(y) = (a^2 + b^2)(c^2 + d^2) = a^2c^2 + b^2d^2 + b^2c^2 + a^2d^2.$$

9. Suppose $a = bu$, where u is a unit. Then $d(b) \leq d(bu) = d(a)$. Also, $d(a) \leq d(au^{-1}) = d(b)$.

11. $m = 0$ and $n = -1$ give $q = -i, r = -2 - 2i$.

13. First observe that $21 = 3 \cdot 7$ and that $21 = (1 + 2\sqrt{-5})(1 - 2\sqrt{-5})$. To prove that 3 is irreducible in $Z[\sqrt{-5}]$ suppose that $3 = xy$, where $x, y \in Z[\sqrt{-5}]$ and x and y are not units. Then $9 = N(3) = N(x)N(y)$ and, therefore, $N(x) = N(y) = 3$. But there are no integers a and b such that $a^2 + 5b^2 = 3$. The same argument shows that 7 is irreducible over $Z[\sqrt{-5}]$. To show that $1 + 2\sqrt{-5}$ is irreducible over $Z[\sqrt{-5}]$ suppose that $1 + 2\sqrt{-5} = xy$, where $x, y \in Z[\sqrt{-5}]$ and x and y are not units. Then $21 = N(1 + 2\sqrt{-5}) = N(x)N(y)$. Thus $N(x) = 3$ or $N(x) = 7$, both of which are impossible.

15. First observe that $10 = 2 \cdot 5$ and that $10 = (2 - \sqrt{-6})(2 + \sqrt{-6})$. To see that 2 is irreducible over $Z[\sqrt{-6}]$ assume that $2 = xy$, where $x, y \in Z[\sqrt{-6}]$ and x and y are not units. Then $4 = N(2) = N(x)N(y)$ so that $N(x) = 2$. But 2 cannot be written in the form $a + 6b^2$. A similar argument applies to 5. To see that $2 - \sqrt{-6}$ is irreducible suppose that $2 - \sqrt{-6} = xy$ where $x, y \in Z[\sqrt{-6}]$ and x and y are not units. Then $10 = N(2 - \sqrt{-6}) = N(x)N(y)$ and as before this is impossible. We know that $Z[\sqrt{-6}]$ is not a principle ideal domain because a PID is a UFD (Theorem 18.3).

17. Suppose $3 = \alpha\beta$, where $\alpha, \beta \in Z[i]$ and neither is a unit. Then $9 = d(3) = d(\alpha)d(\beta)$, so that $d(\alpha) = 3$. But there are no integers such that $a^2 + b^2 = 3$. Observe that $2 = -i(1 + i)^2$ and $5 = (1 + 2i)(1 - 2i)$ and $1 + i, 1 + 2i$, and $1 - 2i$ are not units.

19. Use Exercise 1 with $d = -1$. 5 and $1 + 2i$; 13 and $3 + 2i$; 17 and $4 + i$.

21. Suppose that $1 + 3\sqrt{-5} = xy$, where $x, y \in Z[\sqrt{-5}]$ and x and y are not units. Then $10 = N(1 + 3\sqrt{-5}) = N(x)N(y)$. Thus, $N(x) = 2$ or $N(x) = 5$. But neither 2 nor 5 can be written in the form $a^2 + 5b^2$ so $1 + 3\sqrt{-5}$ is irreducible over $Z[\sqrt{-5}]$. To see that $1 + 3\sqrt{-5}$ is not prime, observe that $(1 + 3\sqrt{-5})(1 - 3\sqrt{-5}) = 1 + 45 = 46$ so that $1 = 3\sqrt{-5}$ divides $2 \cdot 23$. For $1 + 3\sqrt{-5}$ to divide 2 we need $46 = N(1 + 3\sqrt{-5})$ divides $N(2) = 4$. Likewise, for $1 + 3\sqrt{-5}$ to divide 23 we need that 46 divides 23^2. Since neither of these is true, $1 + 3\sqrt{-5}$ is not prime.

23. First observe that $(-1 + \sqrt{5})(1 + \sqrt{5}) = 4 = 2 \cdot 2$ and by Exercise 22, $1 + \sqrt{5}$ and 2 are irreducible over $Z[\sqrt{5}]$. To see that $-1 + \sqrt{5}$ is irreducible over $Z[\sqrt{5}]$ suppose that $-1 + \sqrt{5} = xy$

where $x, y \in Z[\sqrt{5}]$ and x and y are not units. Let $x = a + b\sqrt{5}$. Then $4 = N(-1 + \sqrt{5}) = N(x)N(y)$ so that $a^2 - 5b^2 = \pm 2$. Viewing this equation modulo 5 gives us $a^2 = 2$ or $a^2 = -2 = 3$. However, every square in Z_5 is 0, 1, or 4.

25. Suppose that $x = a + b\sqrt{d}$ is a unit in $Z[\sqrt{d}]$. Then $1 = N(x) = a^2 + (-d)b^2$. But $-d > 1$ implies that $b = 0$ and $a = \pm 1$.

27. See Example 3.

29. Note that $(a_1 a_2 \cdots a_{n-1})a_n \in \langle p \rangle$ implies $a_1 a_2 \cdots a_{n-1} \in \langle p \rangle$ or $a_n \in \langle p \rangle$. Thus, by induction, p divides some a_i.

31. By Exercise 10, $\langle p \rangle$ is maximal and by Theorem 14.4, $D/\langle p \rangle$ is a field.

33. Suppose R satisfies the ascending chain condition and there is an ideal I of R that is not finitely generated. Then pick $a_1 \in I$. Since I is not finitely generated, $\langle a_1 \rangle$ is a proper subset of I, so we may choose $a_2 \in I$ but $a_2 \notin \langle a_1 \rangle$. As before, $\langle a_1, a_2 \rangle$ is proper, so we may choose $a_3 \in I$ but $a_3 \notin \langle a_1, a_2 \rangle$. Continuing in this fashion, we obtain a chain of infinite length $\langle a_1 \rangle \subset \langle a_1, a_2 \rangle \subset \langle a_1, a_2, a_3 \rangle \subset \cdots$.

Now suppose every ideal of R is finitely generated and there is a chain $I_1 \subset I_2 \subset I_3 \subset \cdots$. Let $I = \cup I_i$. Then $I = \langle a_1, a_2, \cdots, a_n \rangle$ for some choice of $a_1, a_2, \ldots, a_n$. Since $I = \cup I_i$, each a_i belongs to some member of the union, say $I_{i'}$. Letting $k = \max \{i' \mid i = 1, \ldots, n\}$, we see that all $a_i \in I_k$. Thus, $I \subseteq I_k$ and the chain has length at most k.

35. Say $I = \langle a + bi \rangle$. Then $a^2 + b^2 + I = (a + bi)(a - bi) + I = I$ and therefore $a^2 + b^2 \in I$. For any $c, d \in Z$, let $c = q_1(a^2 + b^2) + r_1$ and $d = q_2(a^2 + b^2) + r_2$, where $0 \le r_1, r_2 < a^2 + b^2$. Then $c + di + I = r_1 + r_2 i + I$.

37. $N(6 + 2\sqrt{-7}) = 64 = N(1 + 3\sqrt{-7})$. The other part follows directly from Exercise 25.

39. Theorem 18.1 shows that primes are irreducible. So, assume that a is an irreducible in a UFD R and that $a|bc$ in R. We must show that $a|b$ or $a|c$. Since $a|bc$, there is an element d in R such that $bc = ad$. Now replacing b, c, and d by their factorizations as a

product of irreducibles, we have by the uniqueness property that a (or an associate of a) is one of the irreducibles in the factorization of bc. Thus, a is a factor of b or a is a factor of c.

SUPPLEMENTARY EXERCISES FOR CHAPTERS 15-18

1. Let ϕ be a ring homomorphism and let Ker $\phi = nZ$. Then by Theorem 15.3 we have $Z/nZ \approx Z_n$ is a field. From Theorem 14.3 we have that nZ is a prime ideal of Z, and from Exercise 8 in Chapter 14, we know that n is a prime.

3. Let $x, y \in A \cap B$ and $r \in A$. The $x - y \in A \cap B$ because $A \cap B$ is a subgroup of A. Also, rx and xr are in A since A is closed under multiplication and rx and xr are in B since B is an ideal. This proves that $A \cap B$ is an ideal of A.

   To prove that second statement, define a mapping from A to $(A + B)/B$ by $\phi(a) = a + B$. Since for any $b \in B$ we have $a + B = a + b + B$ we know that ϕ is onto. ϕ preserves addition and multiplication because
   $\phi(a + a') = a + a' + B = (a + B) + (a' + b) = \phi(a) + \phi(a')$ and
   $\phi(aa') = aa' + B = (a + B)(a' + B) = \phi(a)\phi(a')$. So, ϕ is a ring homomorphism. Now observe that $\phi(a) = a + B = 0 + B$ if and only if $a \in A \cap B$. So, by the First Isomorphism Theorem (Theorem 15.3), $A/(A + B) \approx (A + B)/B$.

5. Define a mapping from $F[x]$ onto $F[x]/\langle f(x) \rangle \oplus F[x]/\langle g(x) \rangle$ by $\phi(h(x)) = (h(x) + \langle f(x) \rangle, h(x) + \langle g(x) \rangle)$. To verify that ϕ is a ring homomorphism note that
   $\phi(h(x) + k(x)) = (h(x) + k(x) + \langle f(x) \rangle, h(x) + k(x) + \langle g(x) \rangle) = (h(x) + \langle f(x) \rangle + k(x) + \langle f(x) \rangle, h(x) + \langle g(x) \rangle + k(x) + \langle g(x) \rangle) = (h(x) + \langle f(x) \rangle, h(x) + \langle g(x) \rangle) + (k(x) + \langle f(x) \rangle, k(x) + \langle g(x) \rangle) = \phi(h(x)) + \phi(k(x))$ and
   $\phi((h(x)k(x)) = (h(x)k(x) + \langle f(x) \rangle, h(x)k(x) + \langle g(x) \rangle) = (h(x) + \langle f(x) \rangle, h(x) + \langle g(x) \rangle)(k(x) + \langle f(x) \rangle, k(x) + \langle g(x) \rangle) = \phi(h(x))\phi(k(x))$.
   Next observe that $h(x)$ belongs to Ker ϕ if and only if $h(x) \in \langle f(x) \rangle$ and $h(x) \in \langle g(x) \rangle$. This condition is equivalent to $f(x)$ divides $h(x)$ and $g(x)$ divides $h(x)$. Since $f(x)$ and $g(x)$ are irreducibles and not associates this is equivalent to $f(x)g(x)$ divides $h(x)$. So, Ker $\phi = \langle f(x)g(x) \rangle$ and the statement follows from the First Isomorphism Theorem (Theorem 15.3).

7. The homomorphism from $Z[x]$ onto $Z_2[x]$ whereby each coefficient is reduced modulo 2 has kernel K equal to the set of all polynomials whose coefficients are even, so K is an ideal. To show that K is prime it suffices to observe that, by Theorem 16.3, $Z_2[x]$ is an integral domain and appeal to Theorem 14.3.

9. As in Example 7 of Chapter 6, the mapping is onto, is one-to-one, and preserves multiplication. Also, $a(x + y)a^{-1} = axa^{-1} + aya^{-1}$, so that it preserves addition as well.

11. We can do both parts by showing that
$$Z[i]/\langle 2+i \rangle = \{0 + \langle 2+i \rangle, 1 + \langle 2+i \rangle, 2 + \langle 2+i \rangle, 3 + \langle 2+i \rangle, 4 + \langle 2+i \rangle\},$$
which is obviously isomorphic to Z_5. To this end note that
$2 + i + \langle 2 + i \rangle = 0 + \langle 2 + i \rangle$ implies that $i + \langle 2 + i \rangle = -2 + \langle 2 + i \rangle$.
Thus, any coset $a + bi + \langle 2 + i \rangle$ can be written in the form
$c + \langle 2 + i \rangle$. Moreover, because
$5 + \langle 2 + i \rangle = (2 + i)(2 - i) + \langle 2 + i \rangle = 0 + \langle 2 + i \rangle$, we may assume
that $0 \leq c < 5$. Lastly, we must prove that these five cosets are
distinct. It is enough to show that if $c = 1, 2, 3$, or 4, then
$c + \langle 2 + i \rangle \neq 0 + \langle 2 + i \rangle$. If $c + \langle 2 + i \rangle = 0 + \langle 2 + i \rangle$ then there is an
element $a + bi$ such that $c = (2 + i)(a + bi)$. But then
$c^2 = N(c) = N(2 + i)N(a + bi) = 5(a^2 + b^2)$, which implies that 5
divides c.

13. Observe that $(3 + 2\sqrt{2})(3 - 2\sqrt{2}) = 1$. So,
$(3 + 2\sqrt{2})^n (3 - 2\sqrt{2})^n = ((3 + 2\sqrt{2})(3 - 2\sqrt{2}))^n = 1$.

15. In Z_n we are given $(k + 1)^2 = k + 1$. So, $k^2 + 2k + 1 = k + 1$ or
$k^2 = -k = n - k$. Also, $(n - k)^2 = n^2 - 2nk + k^2 = k^2$, so
$(n - k)^2 = n - k$.

17. If there is a solution $x^2 + y^2 = 2003$ using integers then by reducing
modulo 4, there is a solution to $x^2 + y^2 = 3$ in Z_4. But the only
squares in Z_4 are 0 and 1.

19. By the Mod 2 Irreducibility Test (Theorem 17.3 with $p = 2$) it is
enough to show that $x^4 + x^3 + 1$ is irreducible over Z_2. By
inspection, $x^4 + x^3 + 1$ has no zeros in Z_2 and so it has no linear
factors over Z_2. The only quadratic irreducibles in $Z_2[x]$ are
$x^2 + x + 1$ and $x^2 + 1$ and these are ruled out as factors by long
division.

21. By Theorem 14.4, we can do both parts by showing that $Z[\sqrt{2}]/\langle 2 \rangle$ has only two elements. First note that the coset $a + b\sqrt{2} + \langle \sqrt{2} \rangle = a + \sqrt{2}$. Moreover, since $2 + \langle \sqrt{2} \rangle = (\sqrt{2} + \langle \sqrt{2} \rangle)(\sqrt{2} + \langle \sqrt{2} \rangle) = (0 + \langle \sqrt{2} \rangle)(0 + \langle \sqrt{2} \rangle) = (0 + \langle \sqrt{2} \rangle)$ we can write every coset in $Z[\sqrt{2}]/\langle \sqrt{2} \rangle$ in the form $2k + \langle \sqrt{2} \rangle = 0 + \langle \sqrt{2} \rangle$ or $2k + 1 + \langle \sqrt{2} \rangle = 1 + \langle \sqrt{2} \rangle$.

23. Because $Z[i]/\langle 3 \rangle \approx Z_3[i]$ (map $a + bi$ to $a \bmod 3 + (b \bmod 3)i$ and use Theorem 15.3) is a field (see Example 9 in Chapter 13) we know from Theorem 14.4 that $\langle 3 \rangle$ is maximal.

25. Say $a/b, c/d \in R$. Then $a/b - c/d = (ad - bc)/(bd)$ and $ac/(bd) \in R$ by Euclid's Lemma. By Corollary 3 Theorem 15.5, the field of quotients is Q.

27. Since $Z[i]/\langle 3 \rangle \approx Z_3[i]$ it is a field (see Exercise 23). But $(1,0)(0,1) = (0,0)$ shows that $Z_3 \oplus Z_3$ is not a field.

29. Define a mapping from $R[x]$ to $(R/I)[x]$ by
$\phi(a_n x^n + \cdots + a_0) = (a_n + I)x^n + \cdots + (a_0 + I)$. Since
$\phi((a_n x^n + \cdots + a_0) + (b_n x^n + \cdots + b_0)) =$
$\phi((a_n + b_n)x^n + \cdots + (a_0 + b_0)) = (a_n + b_n + I)x^n + \cdots + (a_0 + b_0 + I) =$
$(a_n + I)x^n + \cdots + (a_0 + I) + (b_n + I)x^n + \cdots + (b_0 + I) =$
$\phi(a_n x^n + \cdots + a_0) + \phi(b_n x^n + \cdots + b_0)$ and
$\phi((a_n x^n + \cdots + a_0)(b_n x^n + \cdots + b_0)) = \phi((a_n b_n)x^n + \cdots + (a_0 b_0)) =$
$(a_n b_n + I)x^n + \cdots + (a_0 b_0 + I) = ((a_n + I)x^n + \cdots + (a_0 + I))((b_n + I)x^n + \cdots + (b_0 + I)) = \phi(a_n x^n + \cdots + a_0)\phi(b_n x^n + \cdots + b_0)$, ϕ is a ring homomorphism. Next note that $\phi(a_n x^n + \cdots + a_0) = 0 + I$ if and only if $a_n, \ldots, a_0 \in I$. So, by Theorem 15.3,
$R[x]/I[x] \approx (R/I)[x]$.

31. Let $I = \langle 2 \rangle$. Then $Z_8[x]/I$ is isomorphic to $Z_2[x]$. (The mapping that takes $a_n x^n + \cdots a_0$ to $(a_n \bmod 2)x^n + \cdots + a_0 \bmod 2$ is a ring homomorphism with kernel I.) Since $Z_2[x]$ is an integral domain (see Theorem 16.3) but not a field ($f(x) = x$ does not an inverse), the same is true for $Z_8[x]/I$.

33. Let $I = \langle x, 3 \rangle$. Then any element $a_n x^n + \cdots + a_0 + I$ of $Z[x]/I$ simplifies to $a_0 \bmod 3$ (since I absorbs all terms with an x and all multiples of 3). So, $Z[x]/I = \{0 + I, 1 + I, 2 + I\}$.

CHAPTER 19

Vector Spaces

1. Each of the four sets is an Abelian group under addition. The verification of the four conditions involving scalar multiplication is straight forward. $\mathbf{R}^n$ has basis
   $\{(1,0,\ldots,0),(0,1,0,\ldots,0),\ldots,(0,0,\ldots,1)\}$;

   $M_2(Q)$ has basis $\left\{\begin{bmatrix} 1 & 0 \\ 0 & 0 \end{bmatrix}, \begin{bmatrix} 0 & 1 \\ 0 & 0 \end{bmatrix}, \begin{bmatrix} 0 & 0 \\ 1 & 0 \end{bmatrix}, \begin{bmatrix} 0 & 0 \\ 0 & 1 \end{bmatrix}\right\}$;

   $Z_p[x]$ has basis $\{1, x, x^2, \ldots\}$; $\mathbf{C}$ has basis $\{1, i\}$.

3. $(a_2x^2 + a_1x + a_0) + (a_2'x^2 + a_1'x + a_0') = (a_2 + a_2')x^2 + (a_1 + a_1')x + (a_0 + a_0')$
   and $a(a_2x^2 + a_1x + a_0) = aa_2x^2 + aa_1x + aa_0$. A basis is $\{1, x, x^2\}$.
   Yes, this set $\{x^2 + x + 1, x + 5, 3\}$ is a basis because
   $a(x^2 + x + 1) + b(x + 5) = 3c = 0$ implies that
   $ax^2 + (a + b)x + a + 5b + 3c = 0$. So, $a = 0, a + b = o$ and
   $a + 5b + 3c = 0$. But the two conditions $a = 0$ and $a + b = 0$ imply
   $b = 0$ and the three conditions $a = 0, b = 0$, and $a + 5b + 3c = 0$
   imply $c = 0$.

5. They are linearly dependent, since
   $-3(2, -1, 0) - (1, 2, 5) + (7, -1, 5) = (0, 0, 0)$.

7. Suppose $au + b(u + v) + c(u + v + w) = 0$. Then
   $(a + b + c)u + (b + c)v + cw = 0$. Since $\{u, v, w\}$ are linearly
   independent, we obtain $c = 0, b + c = 0$, and $a + b + c = 0$. So,
   $a = b = c = 0$.

9. If the set is linearly independent, it is a basis. If not, then delete
   one of the vectors that is a linear combination of the others (see
   Exercise 8). This new set still spans V. Repeat this process until
   you obtain a linearly independent subset. This subset will still span
   V since you deleted only vectors that are linear combinations of the
   remaining ones.

11. Let u_1, u_2, u_3 be a basis for U and w_1, w_2, w_3 be a basis for W.
    Since dim $V = 5$, there must be elements $a_1, a_2, a_3, a_4, a_5, a_6$ in F,

not all 0, such that $a_1u_1 + a_2u_2 + a_3u_3 + a_4w_1 + a_5w_2 + a_6w_3 = 0$.
Then $a_1u_1 + a_2u_2 + a_3u_3 = -a_4w_1 - a_5w_2 - a_6w_3$ belongs to $U \cap W$
and this element is not 0 because that would imply that
a_1, a_2, a_3, a_4, a_5 and a_6 are all 0.

In general, if dim U + dim W > dim V, then $U \cap W \neq \{0\}$.

13. No. x^2 and $-x^2 + x$ belong to V but their sum does not.

15. Yes, W is a subspace. If (a, b, c) and (a', b', c') belong to W then
$a + b = c$ and $a' + b' = c'$. Thus,
$a + a' + b + b' = (a + b) + (a' + b') = c + c'$ so $(a, b, c) + (a' + b' + c')$
belongs to W and therefore W is closed addition. Also, if (a, b, c)
belongs to W and d is a real number then $d(a, b, c) = (da, db, dc)$
and $ad + bd = cd$ so W is closed under scalar multiplication.

17. $\begin{bmatrix} a & a+b \\ a+b & b \end{bmatrix} + \begin{bmatrix} a' & a'+b' \\ a'+b' & b' \end{bmatrix} =$
$\begin{bmatrix} a+a' & a+b+a'+b' \\ a+b+a'+b' & b+b' \end{bmatrix}$ and
$c \begin{bmatrix} a & a+b \\ a+b & b \end{bmatrix} = \begin{bmatrix} ac & ac+bc \\ ac+bc & bc \end{bmatrix}$.

19. Suppose B is a basis. Then every member of V is some linear
combination of elements of B. If
$a_1v_1 + \cdots + a_nv_n = a_1'v_1 + \cdots + a_n'v_n$, where $v_i \in B$, then
$(a_1 - a_1')v_1 + \cdots + (a_n - a_n')v_n = 0$ and $a_i - a_i' = 0$ for all i.
Conversely, if every member of V is a unique linear combination of
elements of B, certainly B spans V. Also, if $a_1v_1 + \cdots + a_nv_n = 0$,
then $a_1v_1 + \cdots + a_nv_n = 0v_1 + \cdots + 0v_n$ and therefore $a_i = 0$ for all
i.

21. Since $w_1 = a_1u_1 + a_2u_2 + \cdots + a_nu_n$ and $a_1 \neq 0$, we have
$u_1 = a_1^{-1}(w_1 - a_2u_2 - \cdots - a_nu_n)$, and therefore
$u_1 \in \langle w_1, u_2, \ldots, u_n \rangle$. Clearly, $u_2, \ldots, u_n \in \langle w_1, u_2, \ldots, u_n \rangle$. Hence
every linear combination of $u_1, \ldots, u_n$ is in $\langle w_1, u_2, \ldots, u_n \rangle$.

23. Since $(a, b, c, d) = (a, b, a, a+b) = a(1, 0, 1, 1) + b(0, 1, 0, 1)$ and
$(1, 0, 1, 1)$ and $(0, 1, 0, 1)$ are linearly independent, these two vectors
are a basis.

25. Suppose that $B_1 = \{u_1, u_2, \ldots, u_n\}$ is a finite basis for V and B_2 is
an infinite basis for V. Let $w_1, w_2, \ldots, w_{n+1}$ be distinct elements of

B_2. Then, as in the proof of Theorem 19.1, the set $\{w_1, w_2, \ldots, w_n\}$ spans V. This means that w_{n+1} is a linear combination of $w_1, w_2, \ldots, w_n$. But then B_2 is not a linearly independent set.

27. If V and W are vector spaces over F, then the mapping must preserve addition and scalar multiplication. That is, $T : V \to W$ must satisfy $T(u + v) = T(u) + T(v)$ for all vectors u and v in V, and $T(au) = aT(u)$ for all vectors u in V and all scalars a in F. A vector space isomorphism from V to W is a one-to-one linear transformation from V onto W.

29. Suppose v and u belong to the kernel and a is a scalar. Then
$T(v + u) = T(v) + T(u) = 0 + 0 = 0$ and $T(av) = aT(u) = a \cdot 0 = 0.$

31. Let $\{v_1, v_2, \ldots, v_n\}$ be a basis for V. The mapping given by $\phi(a_1 v_1 + a_2 v_2 + \cdots + a_n v_n) = (a_1, a_2, \ldots, a_n)$ is a vector space isomorphism. By observation, ϕ is onto. ϕ is one-to-one because $(a_1, a_2, \ldots, a_n) = (b_1, b_2, \ldots, b_n)$ implies that $a_1 = b_1, a_2 = b_2, \ldots, a_n = b_n$. Since
$\phi((a_1 v_1 + a_2 v_2 + \cdots + a_n v_n) + (b_1 v_1 + b_2 v_2 + \cdots + b_n v_n)) =$
$\phi((a_1 + b_1)v_1 + (a_2 + b_2)v_2 + \cdots + (a_n + b_n)v_n)) =$
$(a_1 + b_1, a_2 + b_2, \ldots, a_n + b_n) = (a_1, a_2, \ldots, a_n) + (b_1, b_2, \ldots, b_n) =$
$\phi((a_1 v_1 + a_2 v_2 + \cdots + a_n v_n) + \phi((b_1 v_1 + b_2 v_2 + \cdots + b_n v_n))$ we have shown that ϕ preserves addition. Moreover, for any c in F we have
$\phi(c(a_1 v_1 + a_2 v_2 + \cdots + a_n v_n)) = \phi(ca_1 v_1 + ca_2 v_2 + \cdots + ca_n v_n)) =$
$(ca_1, ca_2, \ldots, ca_n) = c(a_1, a_2, \ldots, a_n) = c\phi((a_1 v_1 + a_2 v_2 + \cdots + a_n v_n))$
so that ϕ also preserves scalar multiplication.

CHAPTER 20
Extension Fields

1. $\{a5^{2/3} + b5^{1/3} + c \mid a, b, c \in Q\}$.

3. Since $x^3 - 1 = (x - 1)(x^2 + x + 1)$ the zeros of $x^3 - 1$ are $1, (-1 + \sqrt{-3})/2$, and $(-1 - \sqrt{-3})/2$. So, the splitting field is $Q(\sqrt{-3})$.

5. Since the zeros of $x^2 + x + 1$ are $(-1 \pm \sqrt{-3})/2$ and the zeros of $x^2 - x + 1$ are $(1 \pm \sqrt{-3})/2$, the splitting field is $Q(\sqrt{-3})$.

7. Note that $a = \sqrt{1 + \sqrt{5}}$ implies that $a^4 - 2a^2 - 4 = 0$. Then $p(x) = x^4 - 2x^2 - 4$ is irreducible over Q (to see this let $y = x^2$ and apply the quadratic formula to $y^2 - 2y - 4$) and $p(a) = 0$. So, by Theorem 20.3, $Q(\sqrt{1 + \sqrt{5}})$ is isomorphic to $Q[x]/\langle p(x)\rangle$.

9. Since $a^3 + a + 1 = 0$ we have $a^3 = a + 1$. Thus, $a^4 = a^2 + a; a^5 = a^3 + a^2 = a^2 + a + 1$. To compute a^{-2} and a^{100}, we observe that $a^7 = 1$, since $F(a)^*$ is a group of order 7. Thus, $a^{-2} = a^5 = a^2 + a + 1$ and $a^{100} = (a^7)^{14}a^2 = a^2$.

11. $Q(\pi)$ is the set of all expressions of the form

$$(a_n\pi^n + a_{n-1}\pi^{n-1} + \cdots + a_0)/(b_m\pi^m + b_{m-1}\pi^{m-1} + \cdots + b_0),$$

where $b_m \neq 0$.

13. $x^7 - x = x(x^6 - 1) = x(x^3 + 1)(x^3 - 1) = x(x - 1)^3(x + 1)^3; x^{10} - x = x(x^9 - 1) = x(x - 1)^9$ (see Exercise 41 of Chapter 13).

15. Let b be a zero of $f(x)$ in some extension of F. Then $b^p = a$ and $f(x) = x^p - b^p = (x - b)^p$ (see Exercise 41 of Chapter 13). So, if $b \in F$ then $f(x)$ splits in F and if $b \notin F$ then $f(x)$ is irreducible over F.

17. Solving $1 + \sqrt[3]{4} = (a + b\sqrt[3]{2} + c\sqrt[3]{4})(2 - 2\sqrt[3]{2})$ for a, b, and c yields $a = 4/3, b = 2/3$, and $c = 5/6$.

19. Since $1 + i = -(4 - i) + 5$, $Q(1 + i) \subseteq Q(4 - i)$; conversely,
    $4 - i = 5 - (1 + i)$ implies that $Q(4 - i) \subseteq Q(1 + i)$.

21. If the zeros of $f(x)$ are $a_1, a_2, \ldots, a_n$ then the zeros of $f(x + a)$ are
    $a_1 - a, a_2 - a, \ldots a_n - a$. So, by Exercise 20, $f(x)$ and $f(x - a)$ have
    the same splitting field.

23. Clearly, Q and $Q(\sqrt{2})$ are subfields of $Q(\sqrt{2})$. Assume that there is
    a subfield F of $Q(\sqrt{2})$ that contains an element $a + b\sqrt{2}$ with $b \neq 0$.
    Then, since every subfield of $Q(\sqrt{2})$ must contain Q, we have by
    Exercise 20 that $Q(\sqrt{2}) = Q(a + b\sqrt{2}) \subseteq F$. So, $F = Q(\sqrt{2})$.

25. Let $F = Z_3[x]/\langle x^3 + 2x + 1 \rangle$ and denote the coset $x + \langle x^3 + 2x + 1 \rangle$
    by β and the coset $2 + \langle x^3 + 2x + 1 \rangle$ by 2. Then β is a zero of
    $x^3 + 2x + 1$ and therefore $\beta^3 + 2\beta + 1 = 0$. Using long division we
    obtain $x^3 + 2x + 1 = (x - \beta)(x^2 + \beta x + (2 - \beta^2))$. By trial and error
    we discover that $\beta + 1$ is a zero of $x^2 + \beta x + (2 - \beta^2)$ and by long
    division we deduce that $-2\beta - 1$ is the other zero of
    $x^2 + \beta x + (2 - \beta^2)$. So, we have
    $x^3 + 2x + 1 = (x - \beta)((x - \beta - 1)(x + 2\beta + 1)$.

27. Suppose that $\phi : Q(\sqrt{-3}) \rightarrow Q(\sqrt{3})$ is an isomorphism. Since
    $\phi(1) = 1$, we have $\phi(-3) = -3$. Then
    $-3 = \phi(-3) = \phi(\sqrt{-3}\sqrt{-3}) = (\phi(\sqrt{-3}))^2$. This is impossible, since
    $\phi(\sqrt{-3})$ is a real number.

29. By long division we obtain $x^2 + x + 1 = (x - \beta)(x + 1 + \beta)$ so the
    other zero is $-1 - \beta$.

31. Since $f(x) = x^{21} + 2x^8 + 1$ and $f'(x) = x$ have no common factor of
    positive degree we know by Theorem 20.5 that $f(x)$ has no multiple
    zeros in any extension of Z_3.

33. Since $f(x) = x^{p^n} - x$ and $f'(x) = -1$ have no common factor of
    positive degree we know by Theorem 20.5 that $f(x)$ has no multiple
    zeros in any extension of Z_3.

CHAPTER 21
Algebraic Extensions

1. It follows from Theorem 21.1 that if $p(x)$ and $q(x)$ are both monic irreducible polynomials in $F[x]$ with $p(a) = q(a) = 0$, then deg $p(x) =$ deg $q(x)$. If $p(x) \neq q(x)$, then
$(p - q)(a) = p(a) - q(a) = 0$ and deg $(p(x) - q(x)) <$ deg $p(x)$,
contradicting Theorem 21.1.

   To prove Theorem 21.3 we use the Division Algorithm (Theorem 16.2) to write $f(x) = p(x)q(x) + r(x)$, where $r(x) = 0$ or deg $r(x) <$ deg $p(x)$. Since $0 = f(a) = p(a)q(a) + r(a) = r(a)$ and $p(x)$ is a polynomial of minimum degree for which a is a zero, we may conclude that $r(x) = 0$.

3. Let $F = Q(\sqrt{2}, \sqrt[3]{2}, \sqrt[4]{2}, \ldots)$. Since $[F : Q] \geq [Q(\sqrt[n]{2}) : Q] = n$ for all n, $[F : Q]$ is infinite. To prove that F is an algebraic extension of Q, let $a \in F$. There is some k such that $a \in Q(\sqrt{2}, \sqrt[3]{2}, \sqrt[4]{2}, \ldots, \sqrt[k]{2})$. It follows from Theorem 21.5 that $[Q(\sqrt{2}, \sqrt[3]{2}, \sqrt[4]{2}, \ldots, \sqrt[k]{2}) : Q]$ is finite and from Theorem 21.4 that $Q(\sqrt{2}, \sqrt[3]{2}, \sqrt[4]{2}, \ldots, \sqrt[k]{2})$ is algebraic.

5. Since every irreducible polynomial in $F[x]$ is linear, every irreducible polynomial in $F[x]$ splits in F. So, by Exercise 4, F is algebraically closed.

7. Suppose $Q(\sqrt{a}) = Q(\sqrt{b})$. If $\sqrt{b} \in Q$, then $\sqrt{a} \in Q$ and we may take $c = \sqrt{a}/\sqrt{b}$. If $\sqrt{b} \notin Q$, then $\sqrt{a} \notin Q$. Write $\sqrt{a} = r + s\sqrt{b}$ where r and s belong to Q. Then $r = 0$ for, if not, then $a = r^2 + 2rs\sqrt{b} + b$ and therefore $(a - r^2 - b)/2r = s\sqrt{b}$. But $(a - r^2 - b)/2r$ is rational whereas $s\sqrt{b}$ is irrational.

   Conversely, if there is a element $c \in Q$ such that $a = bc^2$ (we may assume that c is positive) then, by Exercise 20 in Chapter 20, $Q(\sqrt{a}) = Q(\sqrt{bc^2}) = Q(c\sqrt{b}) = Q(\sqrt{b})$.

9. Since $[E : F] = [E : F(a)][F(a) : F]$ we have $[F(a) : F] = [E : F]$, in which case $F(a) = E$, or $[F(a) : F] = 1$, in which case $F(a) = F$.

11. Note that $[F(a,b) : F]$ is divisible by both $m = [F(a) : F]$ and $n = [F(b) : F]$ and $[F(a,b) : F] \leq mn$. So, $[F(a,b) : F] = mn$.

13. Since a is a zero of $x^3 - a^3$ over $F(a^3)$, we have $[F(a) : F(a^3)] \leq 3$. For the second part, take $F = Q, a = 1$; $F = Q, a = (-1 + i\sqrt{3})/2; F = Q, a = \sqrt[3]{2}$.

15. Suppose $E_1 \cap E_2 \neq F$. Then $[E_1 : E_1 \cap E_2][E_1 \cap E_2 : F] = [E_1 : F]$ implies $[E_1 : E_1 \cap E_2] = 1$, so that $E_1 = E_1 \cap E_2$. Similarly, $E_2 = E_1 \cap E_2$.

17. Since E must be an algebraic extension of $\mathbf{R}$, we have $E \subseteq \mathbf{C}$ and so $[\mathbf{C} : E][E : \mathbf{R}] = [\mathbf{C} : \mathbf{R}] = 2$. If $[\mathbf{C} : E] = 2$, then $[E : \mathbf{R}] = 1$ and therefore $E = \mathbf{R}$. If $[\mathbf{C} : E] = 1$, then $E = \mathbf{C}$.

19. Let a be a zero of $p(x)$ in some extension of F. First note $[E(a) : E] \leq [F(a) : F] = \deg p(x)$. Then observe that $[E(a) : F(a)][F(a) : F] = [E(a) : F] = [E(a) : E][E : F]$. This implies that $\deg p(x)$ divides $[E(a) : E]$, so that $\deg p(x) = [E(a) : E]$. It now follows from Theorem 20.3 that $p(x)$ is irreducible over E.

21. Suppose that $\alpha + \beta$ and $\alpha\beta$ are algebraic over Q and that $\alpha \geq \beta$. Then $\sqrt{(\alpha + \beta)^2 - 4\alpha\beta} = \sqrt{\alpha^2 - 2\alpha\beta + \beta^2} = \sqrt{(\alpha - \beta)^2} = \alpha - \beta$ is also algebraic over Q. Also, $\alpha = ((\alpha + \beta) - (\alpha - \beta))/2$ is algebraic over Q, which is a contradiction.

23. It follows from the Quadratic Formula that $\sqrt{b^2 - 4ac}$ is a primitive element.

25. By the Factor Theorem (Corollary 2 of Theorem 16.2), we have $f(x) = (x - a)(bx + c)$, where $b, c \in F(a)$. Thus, $f(x) = b(x - a)(x + b^{-1}c)$.

27. Say a is a generator of F^*. Then $F = Z_p(a)$, and it suffices to show that a is algebraic over Z_p. If $a \in Z_p$, we are done. Otherwise, $1 + a = a^k$ for some $k \neq 0$. If $k > 0$, we are done. If $k < 0$, then $a^{-k} + a^{1-k} = 1$ and we are done.

29. If $[K : F] = n$, then there are elements $v_1, v_2, \ldots, v_n$ in K that constitute a basis for K over F. The mapping $a_1v_1 + \cdots + a_nv_n \rightarrow (a_1, \ldots, a_n)$ is a vector space isomorphism from K to F^n. If K is isomorphic to F^n, then the n elements in K

corresponding to $(1, 0, \ldots, 0), (0, 1, \ldots, 0), \ldots, (0, 0, \ldots, 1)$ in F^n constitute a basis for K over F.

31. Observe that $[F(a, b) : F(a)] = [F(a)(b) : F(a)] \leq [F(b) : F] \leq [F(a)(b) : F(b)][F(b) : F] = [F(a)(b) : F] = [F(a, b) : F]$.

33. Note that if $c \in Q(\beta)$ and $c \notin Q$, then
$5 = [Q(\beta) : Q] = [Q(\beta) : Q(c)][Q(c) : Q]$ so that $[Q(c) : Q] = 5$. On the other hand, $[Q(\sqrt{2}) : Q] = 2, [Q(\sqrt[3]{2}) : Q] = 3$, and $[Q(\sqrt[4]{2}) : Q] = 4$.

35. By closure, $Q(\sqrt{a} + \sqrt{b}) \subseteq Q(\sqrt{a}, \sqrt{b})$. Since
$(\sqrt{a} + \sqrt{b})^{-1} = \frac{1}{\sqrt{a}+\sqrt{b}} \frac{\sqrt{a}-\sqrt{b}}{\sqrt{a}-\sqrt{b}} = \frac{\sqrt{a}-\sqrt{b}}{a-b}$ and $a - b \in Q(\sqrt{a} + \sqrt{b})$ we have $\sqrt{a} - \sqrt{b} \in Q(\sqrt{a} + \sqrt{b})$. (The case that $a - b = 0$ is trivial.) It follows that $\sqrt{a} = \frac{1}{2}((\sqrt{a} + \sqrt{b}) + (\sqrt{a} - \sqrt{b}))$ and $\sqrt{b} = \frac{1}{2}((\sqrt{a} + \sqrt{b}) - (\sqrt{a} - \sqrt{b}))$ are in $Q(\sqrt{a}, \sqrt{b})$. So, $Q(\sqrt{a}, \sqrt{b}) \subseteq Q(\sqrt{a} + \sqrt{b})$.

CHAPTER 22
Finite Fields

1. Since $729 = 9^3$, $[GF(729) : GF(9)] = 3$; since $64 = 8^2$, $[GF(64) : GF(8)] = 2$.

3. By Theorem 22.2, there is an element a in K such that $K^* = \langle a \rangle$. Thus $K = F(a)$.

5. The only possibilities for $f(x)$ are $x^3 + x + 1$ and $x^3 + x^2 + 1$. If a is a zero of $x^3 + x + 1$, then $|Z_2(a)| = |Z_2[x]/\langle x^3 + x + 1 \rangle| = 8$. Moreover, testing each of a^2, a^3, a^4 shows that the other two zeros of $x^3 + x + 1$ are a^2 and a^4. So, $Z_2(a)$ is the splitting field for $x^3 + x + 1$.

   For the second case, let a be a zero of $x^3 + x^2 + 1$. As in the first case, $|Z_2(a)| = 8$. Moreover, testing each of a^2, a^3, a^4 shows that the other two zeros are a^2 and a^4. So, $Z_2(a)$ is the splitting field for $x^3 + x^2 + 1$.

7. By Exercise 38 in Chapter 15, ϕ is a ring homomorphism. Since the only ideals of a field are $\{0\}$ and the field itself (Exercise 25 in Chapter 14), Ker $\phi = \{0\}$. Thus, ϕ is an automorphism. To show that ϕ^n is the identity, we first observe by Corollary 4 of Lagrange's Theorem (Theorem 7.1) that $a^{p^n-1} = 1$ for all a in $GF(p^n)^*$. Thus, $\phi^n(a) = a^{p^n} = a^{p^n-1}a = a$ for all a in $GF(p^n)^*$. Obviously, $\phi^n(0) = 0$.

9. If $g(x)$ is an irreducible factor of $x^8 - x$ over Z_2 and deg $g(x) = m$, then the field $Z_2[x]/\langle g(x) \rangle$ has order 2^m and is isomorphic to a subfield of GF(8). So, by Theorem 22.3, $m = 1$ or 3.

11. Since $|(Z_3[x]/\langle x^3 + 2x + 2 \rangle)^*| = 26$, we need only show that $|x| \neq 1, 2$ or 13. Obviously, $x \neq 1$ and $x^2 \neq 1$. Using the fact that $x^3 + 2x + 1 = 0$ and doing the calculations we obtain $x^{13} = 2$. (Or use the computer software for Chapter 22 at www.d.umn.edu/~jgallian.)

13. Direct calculations show that $x^{13} = 1$, whereas $(2x)^2 \neq 1$ and $(2x)^{13} \neq 1$. Thus $2x$ is a generator.

15. Note that if K is any subfield of $GF(p^n)$ then K^* is a subgroup of the cyclic group $GF(p^n)^*$. So, by Theorem 4.3, K^* is the unique subgroup of $GF(p^n)^*$ of its order.

17. Since $x^2 + 1$ has no zeros in Z_3, it is irreducible over Z_3 (see Corollary 2 of Theorem 16.2). By Corollary 1 of Theorem 17.5, $Z_3[x]/\langle x^2 + 1 \rangle$ is a field. Since every element of $Z_3[x]/\langle x^2 + 1 \rangle$ has the form $ax + b + \langle x^2 + 1 \rangle$, the field has order 9. To see the conversion table use the computer software for Chapter 22 at www.d.umn.edu/~jgallian.)

19. Let $a, b \in K$. Then, by Exercise 41b in Chapter 13, $(a - b)^{p^m} = a^{p^m} - b^{p^m} = a - b$. Also, $(ab)^{p^m} = a^{p^m} b^{p^m} = ab$. So, K is a subfield.

21. By Corollary 3 of Lagrange's Theorem (Theorem 7.1), for every element a in F^* we have $a^{p^n - 1} = 1$. So, every element in F^* is a zero of $x^{p^n} - x$.

23. They are identical.

25. The hypothesis implies that $g(x) = x^2 - a$ is irreducible over $GF(p)$. Then a is a square in $GF(p^n)$ if and only if $g(x)$ has a zero in $GF(p^n)$. Since $g(x)$ splits in $GF(p)[x]/\langle g(x) \rangle \approx GF(p^2)$, $g(x)$ has a zero in $GF(p^n)$ if and only if $GF(p^2)$ is a subfield of $GF(p^n)$. The statement now follows from Theorem 22.3.

27. This is a direct consequence of Exercise 7.

29. Since both α^{62} and -1 have order 2 in the cyclic group F^* and a cyclic group of even order has a unique element of order 2 (see Theorem 4.4), we have $\alpha^{62} = -1$.

31. See the solution for Exercise 19.

33. Consider the field of quotients of $Z_p[x]$. The polynomial $f(x) = x$ is not the image of any element.

CHAPTER 23

Geometric Constructions

1.

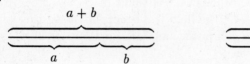

3. Let y denote the length of the hypotenuse of the right triangle with base 1 and x denote the length of the hypotenuse of the right triangle with the base $|c|$. Then $y^2 = 1 + d^2$, $x^2 + y^2 = (1 + |c|)^2$ and $|c|^2 + d^2 = x^2$. So, $1 + 2|c| + |c|^2 = 1 + d^2 + |c|^2 + d^2$, which simplifies to $|c| = d^2$.

5. Suppose that $\sin\theta$ is constructible. Then, by Exercises 1, 2, and 3, $\sqrt{1 - \sin^2\theta} = \cos\theta$ is constructible. Similarly, if $\cos\theta$ is constructible then so is $\sin\theta$.

7. From the identity $\cos 2\theta = 2\cos^2\theta - 1$ we see that $\cos 2\theta$ is constructible if and only if $\cos\theta$ is constructible.

9. By Exercises 5 and 7 to prove that a 45° angle can be trisected it is enough to show that $\sin 15°$ is constructible. To this end note that $\sin 45° = \sqrt{2}/2$ and $\sin 30° = 1/2$ are constructible and $\sin 15° = \sin 45° \cos 30° - \cos 45° \sin 30°$. So, $\sin 15°$ is constructible.

11. Note that solving two linear equations with coefficients in F involves only operations that F is closed under.

13. From Theorem 17.1 and the Rational Root Theorem (Exercise 25 in Chapter 17) it is enough to verify that none of $\pm 1, \pm 1/2, \pm 1/4$, and $\pm 1/8$ is a zero of $8x^3 - 6x - 1$.

15. If a regular 9-gon is constructible then so is the angle $360°/9 = 40°$. But Exercise 10 shows that a 40° angle is not constructible.

17. This amounts to showing $\sqrt{\pi}$ is not constructible. But if $\sqrt{\pi}$ is constructible, so is π. However, $[Q(\pi):Q]$ is infinite.

19. "Tripling" the cube is equivalent to constructing an edge of length $\sqrt[3]{3}$. But $[Q(\sqrt[3]{3}):Q] = 3$, so this can't be done.

21. "Cubing" the circle is equivalent to constructing the length $\sqrt[3]{\pi}$. But $[Q(\sqrt[3]{\pi}):Q]$ is infinite.

SUPPLEMENTARY EXERCISES FOR CHAPTERS 19-23

1. Since $f(x) = x^{50} - 1$ and $f'(x) = 50x^{49}$ have no common factor of positive degree in common we know by Theorem 20.5 that $x^{50} - 1$ has no multiple zeros in any extension of Z_3.

3. Suppose b is one solution of $x^n = a$. Since F^* is a cyclic group of order $q - 1$, it has a cyclic subgroup of order n, say $\langle c \rangle$. Then each member of $\langle c \rangle$ is a solution to the equation $x^n = 1$. It follows that $b\langle c \rangle$ is the solution set of $x^n = a$.

5. $(5a^2 + 2)/a = 5a + 2a^{-1}$. Now observe that since $a^2 + a + 1 = 0$, we know $a(-a - 1) = 1$, so that $a^{-1} = -a - 1$. Thus, $(5a^2 + 2)/a = -2 + 3a$.

7. Since $[F(a) : F] = 5$, $\{1, a, a^2, a^3, a^4\}$ is a basis for $F(a)$ over F. Also, from $5 = [F(a) : F] = [F(a) : F(a^3)][F(a^3) : F]$ we know that $[F(a^3) : F] = 1$ or 5. However, $[F(a^3) : F] = 1$ implies that $a^3 \in F$ and therefore the elements $1, a, a^2, a^3, a^4$ are not linearly independent over F. So, $[F(a^3) : F] = 5$.

9. Since $F(a) = F(a^{-1})$, we have that degree of $a = [F(a) : F] = [F(a^{-1}) : F] = $ degree of a^{-1}.

11. If ab is a zero of $c_n x^n + \cdots + c_1 x + c_0 \in F[x]$, then a is a zero of $c_n b^n x^n + \cdots + c_1 b x + c_0 \in F(b)[x]$.

13. Every element of $F(a)$ can be written in the form $f(a)/g(a)$, where $f(x), g(x) \in F[x]$. If $f(a)/g(a)$ is algebraic and not in F, then there is some $h(x) \in F[x]$ such that $h(f(a)/g(a)) = 0$. By clearing fractions and collecting like powers of a, we obtain a polynomial in a with coefficients from F equal to 0. But then a would be algebraic over F.

15. If K is a finite extension of a finite field F then K itself is a finite field. So, $K^* = \langle a \rangle$ for some $a \in K$ and therefore $K = F(a)$.

17. If the basis elements commute, then so would any combination of basis elements. However, the entire space is not commutative.

19. Since every element in the set has the form $ax^3 + bx^2 + cx$, x, x^2, x^3 is a basis.

21. To prove that L is a subfield we must show that $a, b \in L$ implies that $a - b \in L$ and $ab^{-1} \in L$. To this end, suppose that $a^{p^m} \in F$ and $b^{p^n} \in F$. Then, by Exercise 41 in Chapter 13,
$(a - b)^{p^{m+n}} = a^{p^{m+n}} - b^{p^{m+n}} = (a^{p^m})^{p^n} - (b^{p^n})^{p^m} \in F$. Also,
$(ab^{-1})^{p^{m+n}} = (a^{p^m})^{p^n}((b^{p^n})^{p^m})^{-1} \in F$. So, L is a subfield of K.
Lastly, note that for any $a \in F$ we have $a = a^{p^0}$ so that $a \in L$.

23. By Theorem 20.5, the zeros of $x^n - a$ are distinct, say $\alpha_1, \alpha_2, \ldots, \alpha_n$. Then $\beta_i = \alpha_1/\alpha_i$ for $i = 1, 2, \ldots, n$ are all the nth roots of unity.

CHAPTER 24
Sylow Theorems

1. $a = eae^{-1}$; $cac^{-1} = b$ implies $a = c^{-1}bc = c^{-1}b(c^{-1})^{-1}$; $a = xbx^{-1}$ and $b = ycy^{-1}$ imply $a = xycy^{-1}x^{-1} = xyc(xy)^{-1}$.

3. Observe that $T(xC(a)) = xax^{-1} = yay^{-1} = T(yC(a)) \Leftrightarrow y^{-1}xa = ay^{-1}x \Leftrightarrow y^{-1}x \in C(a) \Leftrightarrow yC(a) = xC(a)$. This proves that T is well defined and one-to-one. Onto is by definition.

5. By way of contradiction, assume that H is the only Sylow 2-subgroup of G and that K is the only Sylow 3-subgroup of G. Then H and K are normal and Abelian (Corollary to Theorem 24.5 and corollary to Theorem 24.2). So, $G = H \times K \approx H \oplus K$ and, from Exercise 4 of Chapter 8, G is Abelian.

7. Suppose that H and K are Sylow 2-subgroups of G. By Lagrange's Theorem (Theorem 7.1), we know that $|H \cap K|$ divides 48. Also, by Exercise 7 in the Supplementary Exercises for Chapters 5–8, we have $|HK| = |H||K|/|H \cap K|$. So, $48 \geq 16 \cdot 16/|H \cap K|$ and therefore $|H \cap K| > 4$.

9. Since $\langle x \rangle$ is a normal subgroup of $N(K)$, by Exercise 51 in Chapter 9, $\langle x \rangle K$ is a subgroup. By Exercise 7 in the Supplementary Exercises for Chapters 5–8, $|\langle x \rangle K| = |\langle x \rangle||K|/|\langle x \rangle \cap K|$, which implies that $|\langle x \rangle K|$ is a power of p. But since $|K|$ is the maximum power of p for any subgroup of G, we have $|\langle x \rangle||K|/|\langle x \rangle \cap K| \leq |K|$. It follows that $|\langle x \rangle|/|\langle x \rangle \cap K| = 1$ and therefore $\langle x \rangle \cap K = \langle x \rangle$.

11. By Theorem 24.5, there are 8 Sylow 7-subgroups.

13. There are two Abelian groups of order 4 and two of order 9. There are both cyclic and dihedral groups of orders 6, 8, 10, 12, and 14. So, 15 is the first candidate. And, in fact, Theorem 24.5 shows that there is only one group of order 15.

15. By Exercise 14, G has seven subgroups of order 3.

17. By Theorem 24.5 the only possibilities are 1, 4, and 10. So, once we find 5 we know there are actually 10. Here are five:
$\langle(123)\rangle, \langle(234)\rangle, \langle(134)\rangle, \langle(345)\rangle, \langle(245)\rangle$.

19. It suffice to show that the correspondence from the set of left cosets of $N(H)$ in G to the set of conjugates of H given by $T(xN(H)) = xHx^{-1}$ is well defined, onto, and one-to-one. Observe that $xN(H) = yN(H) \Leftrightarrow y^{-1}xN(H) = N(H) \Leftrightarrow y^{-1}x \in N(H) \Leftrightarrow y^{-1}xH(y^{-1}x)^{-1} = y^{-1}xHx^{-1}y = H \Leftrightarrow xHx^{-1} = yHy^{-1}$. This shows that T is well defined and one-to-one. By observation, T is onto.

21. Primes are ruled by Corollary 3 of Theorem 7.1; the corollary to Theorem 24.2 rules out 9; Theorem 24.6 rules out 15. This leaves 21.

23. Sylow's Third Theorem (Theorem 24.5) implies that the Sylow 3- and Sylow 5-subgroups are unique. Pick any x not in the union of these. Then $|x| = 15$.

25. By Sylow's Third Theorem, $n_{17} = 1$ or 35. Assume $n_{17} = 35$. Then the union of the Sylow 17-subgroups has 561 elements. By Sylow's Third Theorem, $n_5 = 1$. Thus, we may form a cyclic subgroup of order 85 (Exercise 47 of Chapter 9 and Theorem 24.6). But then there are 64 elements of order 85. This gives too many elements for the group.

27. If $|G| = 60$ and $|Z(G)| = 4$, then by Theorem 24.6, $G/Z(G)$ is cyclic. The "G/Z" Theorem (Theorem 9.3) then tells us that G is Abelian. But if G is Abelian, then $Z(G) = G$.

29. Let H be the Sylow 3-subgroup and suppose that the Sylow 5-subgroups are not normal. By Sylow's Third Theorem, there must be six Sylow 5-subgroups, call them $K_1, \ldots, K_6$. These subgroups have 24 elements of order 5. Also, the cyclic subgroups $HK_1, \ldots, HK_6$ of order 15 each have eight generators. Thus, there are 48 elements of order 15. This gives us more than 60 elements in G.

31. We proceed by induction on $|G|$. By Theorem 24.2 and Theorem 9.5, $Z(G)$ has an element x of order p. By induction, the group $G/\langle x \rangle$ has normal subgroups of order p^k for every k between 1 and $n - 1$, inclusively. By Exercise 49 in Chapter 10, every

subgroup of $G/\langle x \rangle$ has the form $H/\langle x \rangle$, where H is a subgroup of G. Moreover, if $|H/\langle x \rangle| = p^k$ then $|H|$ has order p^{k+1}. So, all that remains is to show is that if $H/\langle x \rangle$ is normal in $G/\langle x \rangle$, then H is normal in G. Let $g \in G$ and $h \in H$. Then
$$g\langle x \rangle h\langle x \rangle ((g\langle x \rangle)^{-1} = g\langle x \rangle h\langle x \rangle g^{-1}\langle x \rangle = ghg^{-1}\langle x \rangle = h'\langle x \rangle \text{ for some}$$
$h' \in H$. So, $ghg^{-1} \in h'\langle x \rangle \subseteq H$.

33. Pick $x \in Z(G)$ such that $|x| = p$. If $x \in H$, by induction, $N(H/\langle x \rangle) > H/\langle x \rangle$, say $y\langle x \rangle \in N(H/\langle x \rangle)$ but $y\langle x \rangle \notin H/\langle x \rangle$. Then y is not in H, and by the argument given in Exercise 31, $y \in N(H)$. If $x \notin H$, then $x \in N(H)$, so that $N(H) > H$.

35. Since automorphisms preserve order, we know $|\alpha(H)| = |H|$. But then the corollary of Theorem 24.5 shows that $\alpha(H) = H$.

37. Since 3 divides $|N(K)|$ we know that $N(K)$ has a subgroup H_1 of order 3. Then, by Exercise 51 in Chapter 9, Exercise 7 of the Supplementary Exercises for Chapters 5–8, and Theorem 24.6, $H_1 K$ is a cyclic group of order 15. Thus, $K \subseteq N(H_1)$ and therefore 5 divides $|N(H_1)|$. And since H and H_1 are conjugates it follows from Exercise 23 in the Supplementary Exercises for Chapters 1–4 that 5 divides $|N(H)|$.

39. Normality of H implies $cl(h) \subseteq H$ for h in H. Now observe that $h \in cl(h)$. This is true only when H is normal.

41. The mapping from H to xHx^{-1} given by $h \to xhx^{-1}$ is an isomorphism.

43. Say $cl(x) = \{x, g_1 x g_1^{-1}, g_2 x g_2^{-1}, \ldots, g_k x g_k^{-1}\}$. If $x^{-1} = g_i x g_i^{-1}$, then for each $g_j x g_j^{-1}$ in $cl(x)$ we have
$(g_j x g_j^{-1})^{-1} = g_j x^{-1} g_j^{-1} = g_j(g_i x g_i^{-1})g_j^{-1} \in cl(x)$. Because $|G|$ has odd order, $g_j x g_j^{-1} \neq (g_j x g_j^{-1})^{-1}$. It follows that $|cl(x)|$ is even. But $|cl(x)|$ divides $|G|$.

45. Let G be a group of order 21. By Sylow's Third Theorem (Theorem 24.5) and its corollary there is a unique Sylow 7-subgroup H, which is normal in G. Let $H = \langle x \rangle$ and let $y \in G$ have order 3. Since $\langle x \rangle$ is normal in G, $yxy^{-1} = x^i$ for some $i = 1, 2, 3, 4, 5,$ or 6. Then $x = y^3 x y^{-3} = y^2(yxy^{-1})y^{-2} = y^2 x^i y^{-2} = y(yx^i y^{-1})y^{-1} = y(yxy^{-1})^i y^{-1} = yx^{i^2} y^{-1} = (yxy^{-1})^{i^2} = (x^i)^{i^2} = x^{i^3}$. So, 7 divides $i^3 - 1$ and we see that the only possibilities for i are $i = 1, 2$ or 4.

47. Say cl(e) and cl(a) are the only two conjugacy classes of a group G of order n. Then cl(a) has $n-1$ elements all of the same order, say m. If $m = 2$, then it follows from Exercise 33 Chapter 2 that G is Abelian. But then cl(a) = $\{a\}$ and so $n = 2$. If $m > 2$, then cl(a) has at most $n-2$ elements since conjugation of a by e, a, and a^2 each yield a.

49. $|C(a)|/|G|$

51. Since D_4 has 5 conjugate classes, $\Pr(D_4) = 5/8$. Since S_3 has 2 conjugacy classes, $\Pr(S_3) = 1/3$. Since A_4 has 4 conjugacy classes, $\Pr(A_4) = 1/3$.

53. Exactly as in the case for a group, we have for a ring $R = \{x_1, x_2, \ldots, x_n\}$, $\Pr(R) = |K|/n^2$, where $K = \{(x,y)|xy = yx, x, y \in R\}$. Also, $|K| = |C(x_1)| + |C(x_2)| + \cdots + |C(x_n)|$. From Exercise 28 in the Supplemental Exercises for Chapters 12–14, we know that $R/C(R)$ is not cyclic. Thus, $|R/C(R)| \geq 4$ and so $|C(R)| \leq |R|/4$. So, for at least 3/4 of the elements x of R, we have $|C(x)| \leq |R|/2$. Then starting with the elements in the center and proceeding to the elements not in the center, we have $|K| \leq |R|/4 + (1/2)(3/4)|R| = (5/8)|R|$.

CHAPTER 25
Finite Simple Groups

1. This follows directly from the "2·odd" Theorem (Theorem 25.2).

3. By the Sylow Theorems if there were a simple group of order 216 the number of Sylow 3-subgroups would be 4. Then the normalizer of a Sylow 3-subgroup would have index 4. The Index Theorem (corollary of Theorem 25.3) then gives a contradiction.

5. Suppose G is a simple group of order 525. Let L_7 be a Sylow 7-subgroup of G. It follows from Sylow's theorems that $|N(L_7)| = 35$. Let L be a subgroup of $N(L_7)$ of order 5. Since $N(L_7)$ is cyclic (Theorem 24.6), $N(L) \geq N(L_7)$, so that 35 divides $|N(L)|$. But L is contained in a Sylow 5-subgroup (Theorem 24.4), which is Abelian (see the Corollary to Theorem 24.2). Thus, 25 divides $|N(L)|$ as well. It follows that 175 divides $|N(L)|$. The Index Theorem now yields a contradiction.

7. Suppose that there is a simple group G of order 528 and L_{11} is a Sylow 11-subgroup. Then $n_{11} = 12, |N(L_{11})| = 44$, and G is isomorphic to a subgroup of A_{12}. Since $|N(L_{11})/C(L_{11})|$ divides $|\text{Aut}(Z_{11})| = 10, |C(L_{11})| = 22$ or 44. In either case, $C(L_{11})$ has elements of order 2 and 11 that commute. But then $C(L_{11})$ has an element of order 22 whereas A_{12} does not.

9. Suppose that there is a simple group G of order 396 and L_{11} is a Sylow 11-subgroup. Then $n_{11} = 12, |N(L_{11})| = 33$, and G is isomorphic to a subgroup of A_{12}. Since $|N(L_{11})/C(L_{11})|$ divides $|\text{Aut}(Z_{11})| = 10, |C(L_{11})| = 33$. Then $C(L_{11})$ has elements of order 3 and 11 that commute. But then $C(L_{11})$ has an element of order 33 whereas A_{12} does not.

11. If we can find a pair of distinct Sylow 2-subgroups A and B such that $|A \cap B| = 8$, then $N(A \cap B) \geq AB$, so that $N(A \cap B) = G$. Now let H and K be any pair of distinct Sylow 2-subgroups. Then $16 \cdot 16/|H \cap K| = |HK| \leq 112$ (Supplementary Exercise 7 for

Chapters 5–8), so that $|H \cap K|$ is at least 4. If $|H \cap K| = 8$, we are done. So, assume $|H \cap K| = 4$. Then $N(H \cap K)$ picks up at least 8 elements from H and at least 8 from K (see Exercise 33 of Chapter 24). Thus, $|N(H \cap K)| \geq 16$ and is divisible by 8. So, $|N(H \cap K)| = 16, 56$, or 112. Since the latter two cases imply that G has a normal subgroup, we may assume $|N(H \cap K)| = 16$. If $N(H \cap K) = H$, then $|H \cap K| = 8$, since $N(H \cap K)$ contains at least 8 elements from K. So, we may assume that $N(H \cap K) \neq H$. Then, we may take $A = N(H \cap K)$ and $B = H$.

15. If A_5 had an element of order 30, 20, or 15, then there would be a subgroup of index 2, 3 or 4. But then the Index Theorem gives us a contradiction to the fact that G is simple.

17. Suppose that there is a simple group of order 120 and let L_3 be a Sylow 3-subgroup of G. By Sylow's Third Theorem (Theorem 24.5), $n_5 = 6$, $n_3 = 10$, or 40. By the Embedding Theorem (Corollary 2 of Theorem 25.3), G is isomorphic to a subgroup of A_6. If $n_3 = 10$, then $|N(L_3)| = 12$ and $|N(L_{11})/C(L_{11})|$ divides $|\text{Aut}(Z_3)| = 2$. So, $C(L_{11})$ has an element of order 2 and an element of order 3 that commute. But then G has an element of order 6 whereas A_6 does not. If $n_3 = 40$, the union of the Sylow 3-subgroups and the Sylow 5-subgroups has 105 elements. We now claim that any two distinct Sylow 2-subgroups L_2 and L_2' have at most 2 elements in common. If this is not the case, then $|L_2 \cap L_2'| = 4$. It then follows from Exercise 33 in Chapter 24 that $N(L_2 \cap L_2')$ contains $L_2 L_2'$. So, by Exercise 7 of the Supplementary Exercises for Chapters 5–8, $N(L_2 \cap L_2') \geq 8 \cdot 8/4 = 16$. We also know that $|N(L_2 \cap L_2')|$ is divisible by 8 and divides 120. So, $|N(L_2 \cap L_2')| \geq 24$. Then, by the Embedding Theorem, G is isomorphic to a subgroup of A_5, a group of order 60. So, L_2 and L_2' have at most 2 elements in common. Thus, when we take the union of three distinct Sylow 2-subgroups (which exist by Sylow's Third Theorem) we produce at least $7 + 6 + 5 = 18$ new elements. This gives us more than 120.

19. Let α be as in the proof of the Generalized Cayley Theorem (Theorem 25.3). Then if $g \in \text{Ker } \alpha$ we have $gH = T_g(H) = H$ so that $\text{Ker } \alpha \subseteq H$. Since $\alpha(G)$ consists of a group of permutations of the left cosets of H in G we know by the First Isomorphism Theorem (Theorem 10.3) that $G/\text{Ker } \alpha$ is isomorphic to a subgroup of $S_{|G:H|}$. Thus, $|G/\text{Ker } \alpha|$ divides $|G : H|!$. Since $\text{Ker } \alpha \subseteq H$, we

have that $|G : H||H : \text{Ker } \alpha| = |G : \text{Ker } \alpha|$ must divide
$|G : H|! = |G : H|(|G : H| - 1)!$. Thus, $|H : \text{Ker } \alpha|$ divides
$(|G : H| - 1)!$. Since $|H|$ and $(|G : H| - 1)!$ are relatively prime, we
have $|H : \text{Ker } \alpha| = 1$ and therefore $H = \text{Ker } \alpha$. So, by the
Corollary of Theorem 10.2, H is normal. In the case that a
subgroup H has index 2, we conclude that H is normal.

21. If H is a proper normal subgroup of S_5, then $H \cap A_5 = A_5$ or $\{\varepsilon\}$
since A_5 is simple and $H \cap A_5$ is normal. But $H \cap A_5 = A_5$ implies
$H = A_5$, whereas $H \cap A_5 = \{\varepsilon\}$ implies $H = \{\varepsilon\}$ or $|H| = 2$. (See
Exercise 19 of Chapter 5.) Now use Exercise 62 of Chapter 9 and
Exercise 46 of Chapter 5.

23. If $PSL(2, Z_7)$ had a nontrivial proper subgroup H, then
$|H| = 2, 3, 4, 6, 7, 8, 12, 14, 21, 24, 28, 42, 56,$ or 84. Observing that
$\begin{bmatrix} 1 & 4 \\ 1 & 5 \end{bmatrix}$ has order 3 and using conjugation we see that $PSL(2, Z_7)$
has more than one Sylow 3-subgroup; observing that $\begin{bmatrix} 5 & 5 \\ 1 & 4 \end{bmatrix}$ has
order 7 and using conjugation we see that $PSL(2, Z_7)$ has more
than one Sylow 7-subgroup; observing that $\begin{bmatrix} 5 & 1 \\ 3 & 5 \end{bmatrix}$ has order 4
and using conjugation we see that $PSL(2, Z_7)$ has more than one
Sylow 2-subgroup. So, from Sylow's Third Theorem, we have
$n_3 = 7, n_7 = 8$, and n_2 is at least 3. So, $PSL(2, Z_7)$ has 14 elements
of order 3, 48 elements of order 7, and at least 11 elements whose
orders are powers of 2. If $|H| = 3, 6,$ or 12, then $|G/H|$ is relatively
prime to 3, and by Exercise 55 of Chapter 9, H would contain the
14 elements of order 3. If $|H| = 24$, then H would contain the 14
elements of order 3 and at least 11 elements whose orders are a
power of 2. If $|H| = 7, 14, 21, 28,$ or 42, then H would contain the
48 elements of order 7. If $|H| = 56$, then H would contain the 48
elements of order 7 and at least 11 elements whose orders are a
power of 2. If $|H| = 84$, then H would contain the 48 elements of
order 7, but by Sylow's Third Theorem a group of order 84 has
only one Sylow 7-subgroup. If $|H| = 2$ or 4, the G/H has a normal
Sylow 7-subgroup. This implies that G would have a normal
subgroup of order 14 or 28, both of which have been ruled out. (To
see that G would have a normal subgroup of order 14 or 28, note
that the natural mapping from G to G/H taking g to gH is a

homomorphism then use properties 8 and 5 of Theorem 10.2.) So, every possibility for H leads to a contradiction.

25. Suppose that S_5 has a subgroup H that contains a 5-cycle α and a 2-cycle β. Say $\beta = (a_1 a_2)$. Then there is some integer k such that $\alpha^k = (a_1 a_2 a_3 a_4 a_5)$. Note that
$(a_1 a_2 a_3 a_4 a_5)^{-1}(a_1 a_2)(a_1 a_2 a_3 a_4 a_5)(a_1 a_2) =$
$(a_5 a_4 a_3 a_2 a_1)(a_1 a_2)(a_1 a_2 a_3 a_4 a_5)(a_1 a_2) = (a_1 a_2 a_5)$, so H contains an element of order 3. Moreover, since
$\alpha^{-2}\beta\alpha^2 = (a_4 a_2 a_5 a_3 a_1)(a_1 a_2)(a_1 a_3 a_5 a_2 a_4) = (a_4 a_5)$, H contains the subgroup $\{(1), (a_1 a_2), (a_4 a_5), (a_1 a_2)(a_4 a_5)\}$. This means that $|H|$ is divisible by 60. But $|H|$ cannot be 60 for if so, then the subset of even permutations in H would be a subgroup of order 30 (see Exercise 19 in Chapter 5). This means that A_5 would have a subgroup of index 2, which would a normal subgroup. This contradicts the simplicity of A_5.

27. Suppose there is a simple group of order 60 that is not isomorphic to A_5. The Index Theorem implies $n_2 \neq 1$ or 3, and the Embedding Theorem implies $n_2 \neq 5$. Thus, $n_2 = 15$. If every pair of Sylow 2-subgroups has only the identity element in common then the union of the 15 Sylow 2-subgroups has 46 elements. But $n_5 = 6$, so there are also 24 elements of order 5. This gives more than 60. As was the case in showing that there is no simple group of order 144 the normalizer of this intersection has index 5, 3, or 1. But the Embedding Theorem and the Index Theorem rule these out.

CHAPTER 26
Generators and Relations

1. $u \sim u$ because u is obtained from itself by no insertions; if v can be obtained from u by inserting or deleting words of the form xx^{-1} or $x^{-1}x$ then u can be obtained from v by reversing the procedure; if u can be obtained from v and v can be obtained from w then u can be obtained from w by first obtaining v from w then u from v.

3.
$$\begin{aligned}
b(a^2N) &= b(aN)a = (ba)Na = a^3bNa = a^3b(aN) = a^3(ba)N \\
&= a^3a^3bN = a^6bN = a^6Nb = a^2Nb = a^2bN \\
b(a^3N) &= b(a^2N)a = a^2bNa = a^2b(aN) = a^2a^3bN \\
&= a^5bN = a^5Nb = aNb = abN \\
b(bN) &= b^2N = N \\
b(abN) &= baNb = a^3bNb = a^3b^2N = a^3N \\
b(a^2bN) &= ba^2Nb = a^2bNb = a^2b^2N = a^2N \\
b(a^3bN) &= ba^3Nb = abNb = ab^2N = aN
\end{aligned}$$

5. Let F be the free group on $\{a_1, a_2, \ldots, a_n\}$. Let N be the smallest normal group containing $\{w_1, w_2, \ldots, w_t\}$ and let M be the smallest normal subgroup containing $\{w_1, w_2, \ldots, w_t, w_{t+1}, \ldots, w_{t+k}\}$. Then $F/N \approx G$ and $F/M \approx \overline{G}$. The homomorphism from F/N to F/M given by $aN \to aM$ induces a homomorphism from G onto $\overline{G}$.

   To prove the corollary, observe that the theorem shows that K is a homomorphic image of G, so that $|K| \le |G|$.

7. Clearly, a and ab belong to $\langle a, b \rangle$, so $\langle a, ab \rangle \subseteq \langle a, b \rangle$. Also, a and $a^{-1}(ab) = b$ belong to $\langle a, ab \rangle$.

9. By Exercise 7, $\langle x, y \rangle = \langle x, xy \rangle$. Also, $(xy)^2 = (xy)(xy) = (xyx)y = y^{-1}y = e$, so by Theorem 26.5, G is isomorphic to a dihedral group and from the proof of Theorem 26.5, $|x(xy)| = |y| = n$ implies that $G \approx D_n$.

11. Since $x^2 = y^2 = e$, we have $(xy)^{-1} = y^{-1}x^{-1} = yx$. Also, $xy = z^{-1}yz$, so that $(xy)^{-1} = (z^{-1}yz)^{-1} = z^{-1}y^{-1}z = z^{-1}yz = xy$.

13. First note that $b^2 = abab$ implies that $b = aba$.

    a. So, $b^2abab^3 = b^2(aba)b^3 = b^2bb^3 = b^6$.

    b. Also, $b^3abab^3a = b^3(aba)b^3a = b^3bb^3a = b^7a$.

15. Note that $yxyx^3 = e$ implies that $yxy^{-1} = x^5$ and therefore $\langle x \rangle$ is normal. So, $G = \langle x \rangle \cup y\langle x \rangle$ and $|G| \le 16$. From $y^2 = e$ and $yxyx^3 = e$, we obtain $yxy^{-1} = x^{-3}$. So, $yx^2y^{-1} = yxy^{-1}yxy^{-1} = x^{-6} = x^2$. Thus, $x^2 \in Z(G)$. On the other hand, G is not Abelian for if so we would have $e = yxyx^3 = x^4$ and then $|G| \le 8$. It now follows from the "G/Z" Theorem (Theorem 9.3) that $|Z(G)| \ne 8$. Thus, $Z(G) = \langle x^2 \rangle$. Finally, $(xy)^2 = xyxy = x(yxy) = xx^{-3} = x^{-2}$, so that $|xy| = 8$.

17. Since the mapping from G onto G/N given by $x \to xN$ is a homomorphism, G/N satisfies the relations defining G.

19. For H to be a normal subgroup we must have $yxy^{-1} \in H = \{e, y^3, y^6, y^9, x, xy^3, xy^6, xy^9\}$. But $yxy^{-1} = yxy^{11} = (yxy)y^{10} = xy^{10}$.

21. First note that $b^{-1}a^2b = (b^{-1}ab)(b^{-1}ab) = a^3a^3 = a^6 = e$. So, $a^2 = e$. Also, $b^{-1}ab = a^3 = a$ implies that a and b commute. Thus, G is generated by an element of order 2 and an element of order 3 that commute. It follows that G is Abelian and has order at most 6. But the defining relations for G are satisfied by Z_6 with $a = 3$ and $b = 2$. So, $G \approx Z_6$.

23. In the notation given in the proof of Theorem 26.5 we have that $|e| = 1$, $|a| = |b| = 2$, $|ab| = |ba| = \infty$. Next observe that since every element of D_∞ can be expressed as a string of alternating a's and b's or alternating b's and a's, every element can be expressed in one of four forms: $(ab)^n, (ba)^n, (ab)^na$, or $(ba)^nb$ for some n. Since $|ab| = |ba| = \infty$, we have $|(ab)^n| = |(ba)^n| = \infty$ (excluding 0). And, since $((ab)^na)^2 = (ab)^na(ab)^na = (ab)(ab)\cdots(ab)a(ab)(ab)\cdots(ab)a$, we can start at the middle and successively cancel the adjacent a's, then adjacent b's, then adjacent a's, and so on to obtain the identity. Thus, $|(ab)^na| = 2$. Similarly, $|(ba)^nb| = 2$.

25. First we show that $d = b^{-1}, a = b^2$ and $c = b^3$ so that $G = \langle b \rangle$. To this end observe that $ab = c$ and $cd = a$ together imply that $cdc = c$ and therefore $d = b^{-1}$. Then $da = b$ and $d = b^{-1}$ together imply

that $a = b^2$. Finally, $cd = a$ and $d = b^{-1}$ together imply $c = b^3$. Thus $G = \langle b \rangle$. Now observe that $bc = d, c = b^3$, and $d = b^{-1}$ yield $b^5 = e$. So $|G| = 1$ or 5. But Z_5 satisfies the defining relations with $a = 1, b = 3, c = 4$, and $d = 2$.

27. There are only five groups of order 8: Z_8 and the quaternions have only one element of order 2; $Z_4 \oplus Z_2$ has 3; $Z_2 \oplus Z_2 \oplus Z_2$ has 7; and D_4 has 5.

CHAPTER 27
Symmetry Groups

1. If T is a distance-preserving function and the distance between points a and b is positive, then the distance between $T(a)$ and $T(b)$ is positive.

3. See Figure 1.5.

5. There are rotations of $0°, 120°$ and $240°$ about an axis through the centers of the triangles and a $180°$ rotation through an axis perpendicular to a rectangular base and passing through the center of the rectangular base. This gives 6 rotations. Each of these can be combined with the reflection plane perpendicular to the base and bisecting the base. So, the order is 12.

7. There are n rotations about an axis through the centers of the n-gons and a $180°$ rotation through an axis perpendicular to a rectangular base and passing through the center of the rectangular base. This gives $2n$ rotations. Each of these can be combined with the reflection plane perpendicular to a rectangular base and bisecting the base. So, the order is $4n$.

9. In $\mathbf{R}^1$, there is the identity and an inversion through the center of the segment. In $\mathbf{R}^2$, there are rotations of $0°$ and $180°$, a reflection across the horizontal line containing the segment, and a reflection across the perpendicular bisector of the segment. In $\mathbf{R}^3$, the symmetry group is $G \oplus Z_2$, where G is the plane symmetry group of a circle. (Think of a sphere with the line segment as a diameter. Then G includes any rotation of that sphere about the diameter and any plane containing the diameter of the sphere is a symmetry in G. The Z_2 must be included because there is also an inversion.)

11. There are 6 elements of order 4 since for each of the three pairs of opposite squares there are rotations of $90°$ and $270°$.

13. An inversion in $\mathbf{R}^3$ leaves only a single point fixed, while a rotation leaves a line fixed.

15. In $\mathbf{R}^4$, a plane is fixed. In $\mathbf{R}^n$, a hyperplane of dimension $n-2$ is fixed.

17. Let T be an isometry, let p, q, and r be the three noncollinear points, and let s be any other point in the plane. Then the quadrilateral determined by $T(p)$, $T(q)$, $T(r)$, and $T(s)$ is congruent to the one formed by p, q, r, and s. Thus, $T(s)$ is uniquely determined by $T(p)$, $T(q)$, and $T(r)$.

19. The center of the rotation is a rotation.

21. The point of intersection of the perpendicular bisector of the line joining p and p and the perpendicular bisector of the line joining q and q.

CHAPTER 28

Frieze Groups and Crystallographic Groups

1. The mapping $\phi(x^m y^n) = (m, n)$ is an isomorphism. Onto is by observation. If $\phi(x^m y^n) = \phi(x^i y^j)$, then $(m, n) = (i, j)$ and therefore, $m = n$ and $i = j$. Also, $\phi((x^m y^n)(x^i y^j)) = \phi(x^{m+i} y^{n+j}) = (m + i, n + j) = (m, n)(i, j) = \phi(x^m y^n)\phi(x^i y^j)$.

3. Using Figure 28.9 we obtain $x^2 y z x z = x y$.

5. Use Figure 28.9.

7. $x^2 y z x z = x^2 y x^{-1} = x^2 x^{-1} y = x y$
   $x^{-3} z x z y = x^{-3} x^{-1} y = x^{-4} y$

9. A subgroup of index 2 is normal.

11. **a.** V, **b.** I, **c.** II, **d.** VI, **e.** VII, and **f.]** III.

13. cmm

15. **a.** $p4m$, **b.** $p3$, **c.** $p31m$, and**d.** $p6m$

17. The principle purpose of tire tread design is to carry water away from the tire. Patterns I and III do not have horizontal reflective symmetry. Thus these designs would not carry water away equally on both halfs of the tire.

19. **a.** VI, **b.** V, **c.** I, **d.** III, **e.** IV, **f.** VII, and **g.** IV

CHAPTER 29
Symmetry and Counting

1. The symmetry group is D_4. Since we have two choices for each vertex, the identity fixes 16 colorings. For R_{90} and R_{270} to fix a coloring, all four corners must have the same color so each of these fixes 2 colorings. For R_{180} to fix a coloring, diagonally opposite vertices must have the same color. So, we have 2 independent choices for coloring the vertices and we can choose 2 colors for each. This gives 4 fixed colorings for R_{180}. For H and V, we can color each of the two vertices on one side of the axis of reflection in 2 ways, giving us 4 fixed points for each of these rotations. For D and D', we can color each of the two fixed vertices with 2 colors and then we are forced to color the remaining two the same. So, this gives us 8 choices for each of these two reflections. Thus, the total number of colorings is

$$\frac{1}{8}(16 + 2 \cdot 2 + 4 + 2 \cdot 4 + 2 \cdot 8) = 6.$$

3. The symmetry group is D_3. There are $5^3 - 5 = 120$ colorings without regard to equivalence. The rotations of 120° and 240° can fix a coloring only if all three vertices of the triangle are colored the same so they each fix 0 colorings. A particular reflection will fix a coloring provided that fixed vertex is any of the 5 colors and the other two vertices have matching colors. This gives $5 \cdot 4 = 20$ for each of the three reflections. So, the number of colorings is

$$\frac{1}{6}(120 + 0 + 0 + 3 \cdot 20) = 30.$$

5. The symmetry group is D_6. The identity fixes all $2^6 = 64$ arrangements. For R_{60} and R_{300}, once we make a choice of a radical for one vertex all others must use the same radical. So, these two fix 2 arrangements each. For R_{120} and R_{240} to fix an arrangement every other vertex must have the same radical. So,

once we select a radical for one vertex and a radical for an adjacent vertex we then have no other choices. So we have 2^2 choices for each of 2 these rotations. For R_{180} to fix an arrangement, each vertex must have the same radical as the vertex diagonally opposite it. Thus, there are 2^3 choices for this case. For the 3 reflections whose axes of symmetry joins two vertices, we have 2 choices for each fixed vertex and 2 choices for each of the two vertices on the same side of the reflection axis. This gives us 16 choices for each of these 3 reflections. For the 3 reflections whose axes of reflection bisects opposite sides of the hexagon, we have 2 choices for each of the 3 vertices on the same size of the reflection axis. This gives us 8 choices for each of these 3 reflections. So, the total number of arrangements is

$$\frac{1}{12}(64 + 2 \cdot 2 + 2 \cdot 4 + 8 + 3 \cdot 16 + 3 \cdot 8) = 13.$$

7. The symmetry group is D_4. The identity fixes $6 \cdot 5 \cdot 4 \cdot 3 = 360$ colorings. All other symmetries fix 0 colorings because of the restriction that no color be used more than once. So, the number of colorings is $360/8 = 45$.

9. The symmetry group is D_{11}. The identity fixes 2^{11} colorings. Each of the other 10 rotations fixes only the two colorings in which the beads are all the same color. (Here we use the fact that 11 is prime. For example, if the rotation $R_{2 \cdot 360/11}$ fixes a coloring then once we choose a color for one vertex the rotation forces all other vertices to have that same color because the rotation moves 2 vertices at a time and 2 is a generator of Z_{11}.) For each reflection, we may color the vertex containing the axis of reflection 2 ways and each vertex on the same side of the axis of reflection 2 ways. This gives us 2^6 colorings for each reflection. So, the number of different colorings is

$$\frac{1}{22}(2^{11} + 10 \cdot 2 + 11 \cdot 2^6) = 126.$$

11. The symmetry group is Z_6. The identity fixes all n^6 possible colorings. Since the rotations of 60° and 300° fix only the cases where each section is the same color, they each fix n colorings. Rotations of 120° and 240° each fix n^2 colorings since every other section must have the same color. The 180° fixes n^3 colorings since

once we choose colors for three adjacent sections the colors for the remaining three sections are determined. So, the number is

$$\frac{1}{6}(n^6 + 2 \cdot n + 2 \cdot n^2 + n^3).$$

13. The first part is Exercise 11 in Chapter 6. For the second part, observe that in D_4 we have $\phi_{R_0} = \phi_{R_{180}}$.

15. We use (12) to denote the permutation that maps L_1 to L_2 and vice versa. Then the mapping that takes rotations to the identity and reflections to (12) is a homomorphism whose kernel is the group of rotations.

CHAPTER 30
Cayley Digraphs of Groups

1. $4^*(b, a)$

3. $(m/2)^*\{3^*[(a,0),(b,0)],(a,0),(e,1),3^*(a,0),(b,0),3^*(a,0),(e,1)\}$

5. $a^3 b$

7. Both yield paths from e to $a^3 b$.

11. Say we start at x. Then we know the vertices
    $x, xs_1, xs_1 s_2, \ldots, xs_1 s_2 \cdots s_{n-1}$ are distinct and $x = xs_1 s_2 \cdots s_n$. So
    if we apply the same sequence beginning at y, then cancellation
    shows that $y, ys_1, ys_1 s_2, \ldots, ys_1 s_2 \cdots s_{n-1}$ are distinct and
    $y = ys_1 s_2 \cdots s_n$.

13. If there were a Hamiltonian path from $(0,0)$ to $(2,0)$, there would
    be a Hamiltonian circuit in the digraph, since $(2,0) + (1,0) = (0,0)$.

15. **a.** If $s_1, s_2, \ldots, s_{n-1}$ traces a Hamiltonian path and
     $s_i s_{i+1} \cdots s_j = e$, then the vertex $s_1 s_2 \cdots s_{i-1}$ appears twice.
     Conversely, if $s_i s_{i+1} \cdots s_j \neq e$, then the sequence
     $e, s_1, s_1 s_2, \ldots, s_1 s_2 \cdots s_{n-1}$ yields the n vertices (otherwise,
     cancellation gives a contradiction).

     b. This is immediate from part a.

17. The sequence traces the digraph in a clockwise fashion.

19. Abbreviate $(a,0)$, $(b,0)$, and $(e,1)$ by a, b, and 1, respectively. A
    circuit is $4^*(4^*1, a), 3^*a, b, 7^*a, 1, b, 3^*a, b, 6^*a, 1, a, b, 3^*a, b, 5^*a,$
    $1, a, a, b, 3^*a, b, 4^*a, 1, 3^*a, b, 3^*a, b, 3^*a, b$.

21. Abbreviate $(R_{90},0)$, $(H,0)$, and $(R_0,1)$ by R, H, and 1, respectively.
    A circuit is $3^*(R, 1, 1), H, 2^*(1, R, R), R, 1, R, R, 1, H, 1, 1$.

23. Abbreviate $(a,0), (b,0)$, and $(e,1)$ by a, b, and 1, respectively. A
    circuit is $2^*(1, 1, a), a, b, 3^*a, 1, b, b, a, b, b, 1, 3^*a, b, a, a$.

25. Abbreviate $(r,0), (f,0)$, and $(e,1)$ by r, f, and 1, respectively. Then the sequence is $r, r, f, r, r, 1, f, r, r, f, r, 1, r, f, r, r, f, 1, r, r,$ $f, r, r, 1, f, r, r, f, r, 1, r, f, r, r, f, 1.$

27. $m^*((n-1)^*(0,1),(1,1))$

29. Abbreviate $(r,0), (f,0)$, and $(e,1)$ by r, f, and 1, respectively. A circuit is $1, r, 1, 1, f, r, 1, r, 1, r, f, 1.$

31. $2*[3*(1,0),(0,1)], 4*(1,0),(0,1), 2*[3*(1,0),(0,1)]$

33. $12*((1,0),(0,1)).$

35. In the proof of Theorem 30.3, we used the hypothesis that G is Abelian in two places: We needed H to satisfy the induction hypothesis, and we needed to form the factor group G/H. Now, if we assume only that G is Hamiltonian, then H also is Hamiltonian and G/H exists.

CHAPTER 31

Introduction to Algebraic Coding Theory

1. wt(000000) = 0; wt(0001011) = 3; wt(0010111) = 4;
   wt(0100101) = 3; wt(1000110) = 3; wt(1100011) = 4;
   wt(1010001) = 3; wt(1001101) = 4; etc.

3. 1000110; 1110100

5. 000000, 100011, 010101, 001110, 110110, 101101, 011011, 111000

7. Not all single errors can be detected.

9. Observe that a vector has even weight if and only if it can be
   written as a sum of an even number of vectors of weight 1. So, if u
   can be written as the sum of $2m$ vectors of even weight and v can
   be written as the sum of $2n$ vectors of even weight, then $u + v$ can
   be written as the sum of $2m + 2n$ vectors of even weight and
   therefore the set of code words of even weight is closed. (We need
   not check that the inverse of a code word is a code word since every
   binary code word is its own inverse.)

11. No, by Theorem 31.3.

13. 0000000, 1000111, 0100101, 0010110, 0001011, 1100010, 1010001,
    1001100, 0110011, 0101110, 0011101, 1110100, 1101001, 1011010,
    0111000, 1111111;

$$H = \begin{bmatrix} 1 & 1 & 1 \\ 1 & 0 & 1 \\ 1 & 1 & 0 \\ 0 & 1 & 1 \\ 1 & 0 & 0 \\ 0 & 1 & 0 \\ 0 & 0 & 1 \end{bmatrix};$$

Yes, the code will detect any single error because it has weight 3.

15. Suppose u is decoded as v and x is the coset leader of the row containing u. Coset decoding means v is at the head of the column containing u. So, $x + v = u$ and $x = u - v$. Now suppose $u - v$ is a coset leader and u is decoded as y. Then y is at the head of the column containing u. Since v is a code word, $u = u - v + v$ is in the row containing $u - v$. Thus $u - v + y = u$ and $y = v$.

17. 000000, 100110, 010011, 001101, 110101, 101011, 011110, 111000

$$H = \begin{bmatrix} 1 & 1 & 0 \\ 0 & 1 & 1 \\ 1 & 0 & 1 \\ 1 & 0 & 0 \\ 0 & 1 & 0 \\ 0 & 0 & 1 \end{bmatrix}$$

001001 is decoded as 001101 by all four methods.
011000 is decoded as 111000 by all four methods.
000110 is decoded as 100110 by all four methods.
Since there are no code words whose distance from 100001 is 1 and three whose distance is 2, the nearest-neighbor method will not decode or will arbitrarily choose a code word; parity-check matrix decoding does not decode 100001; the standard-array and syndrome methods decode 100001 as 000000, 110101, or 101011, depending on which of 100001, 010100, or 001010 is is a coset leader.

19. For any received word w, there are only eight possibilities for wH. But each of these eight possibilities satisfies condition 2 or the first portion of condition 3' of the decoding procedure, so decoding assumes that no error was made or one error was made.

21. There are 3^4 code words and 3^6 possible received words.

23. No; row 3 is twice row 1.

25. No. For if so, nonzero code words would be all words with weight at least 5. But this set is not closed under addition.

27. By Exercise 24, for a linear code to correct every error the minimum weight must be at least 3. Since a (4,2) binary linear code only has three nonzero code words, if each must have weight at least 3 then the only possibilities are (1,1,1,0), (1,1,0,1), (1,0,1,1),(0,1,1,1) and (1,1,1,1). But each pair of these has at least two components that

agree. So, the sum of any distinct two of them is a nonzero word of weight at most 2. This contradicts the closure property.

29. Abbreviate the coset $a + \langle x^2 + x + 1 \rangle$ with a. The following generating matrix will produce the desired code:

$$\begin{bmatrix} 1 & 0 & 1 & 1 & x \\ 0 & 1 & x & x+1 & x+1 \end{bmatrix}.$$

31. By Exercise 14 and the assumption, for each component exactly $n/2$ of the code words have the entry 1. So, determining the sum of the weights of all code words by summing over the contributions made by each component we obtain $n(n/2)$. Thus, the average weight of a code word is $n/2$.

33. Let $c, c' \in C$. Then, $c + (v + c') = v + c + c' \in v + C$ and $(v + c) + (v + c') = c + c' \in C$, so the set $C \cup (v + C)$ is closed under addition.

35. If the ith component of both u and v is 0, then so is the ith component of $u - v$ and au, where a is a scalar.

CHAPTER 32
Introduction to Galois Theory

1. Note that $\phi(1) = 1$. Thus $\phi(n) = n$. Also, for $n \neq 0$, $1 = \phi(1) = \phi(nn^{-1}) = \phi(n)\phi(n^{-1}) = n\phi(n^{-1})$, so that $1/n = \phi(n^{-1})$. So, by properties of automorphisms, $\phi(m/n) = \phi(mn^{-1}) = \phi(m)\phi(n^{-1}) = \phi(m)\phi(n)^{-1} = mn^{-1} = m/n$.

3. If α and β are automorphisms that fix F, then $\alpha\beta$ is an automorphism and, for any x in F, we have $(\alpha\beta)(x) = \alpha(\beta(x)) = \alpha(x) = x$. Also, $\alpha(x) = x$ implies, by definition of an inverse function, that $\alpha^{-1}(x) = x$. So, by the Two-Step Subgroup Test, the set is a group.

5. Suppose that a and b are fixed by every element of H. By Exercise 23 in Chapter 13, it suffices to show that $a - b$ and ab^{-1} are fixed by every element of H. By properties of automorphisms we have for any element ϕ of H, $\phi(a - b) = \phi(a) + \phi(-b) = \phi(a) - \phi(b) = a - b$. Also, $\phi(ab^{-1}) = \phi(a)\phi(b^{-1}) = \phi(a)\phi(b)^{-1} = ab^{-1}$.

7. It suffices to show that each member of $\mathrm{Gal}(K/F)$ defines a permutation on the a_i's. Let $\alpha \in \mathrm{Gal}(K/F)$ and write $f(x) = c_n x^n + c_{n-1} x^{n-1} + \cdots + c_0$. Then $0 = f(a_i) = c_n a_i^n + c_{n-1} a_i^{n-1} + \cdots + c_0$. So, $0 = \alpha(0) = \alpha(c_n)(\alpha(a_i)^n + \alpha(c_{n-1})\alpha(a_i)^{n-1} + \cdots + \alpha(c_0) = c_n(\alpha(a_i))^n + c_{n-1}\alpha(a_i)^{n-1} + \cdots + c_0 = f(\alpha(a_i))$. So, $\alpha(a_i) = a_j$ for some j, and therefore α permutes the a_i's.

9. $\phi^6(\omega) = \omega^{729} = \omega$.
$\phi^3(\omega + \omega^{-1}) = \omega^{27} + \omega^{-27} = \omega^{-1} + \omega$.
$\phi^2(\omega^3 + \omega^5 + \omega^6) = \omega^{27} + \omega^{45} + \omega^{54} = \omega^6 + \omega^3 + \omega^5$.

11. If there were a subfield K of E such that $[K : F] = 2$ then, by the Fundamental Theorem of Galois Theory (Theorem 32.1), A_4 would have a subgroup of index 2. But, by Example 13 in Chapter 9, A_4 has no such subgroup.

13. This follows directly from the Fundamental Theorem of Galois Theory (Theorem 32.1) and Sylow's First Theorem (Theorem 24.3).

15. Let ω be a primitive cube root of 1. Then $Q \subset Q(\sqrt[3]{2}) \subset Q(\omega, \sqrt[3]{2})$ and $Q(\sqrt[3]{2})$ is not the splitting field of a polynomial in $Q[x]$.

17. By the Fundamental Theorem of Finite Abelian Groups (Theorem 11.1), the only Abelian group of order 10 is Z_{10}. By the Fundamental Theorem of Cyclic Groups (Theorem 4.3), the only proper, nontrivial subgroups of Z_{10} are one of index 2 and one of index 5. So, the lattice of subgroups of Z_{10} is a diamond with Z_{10} at the top, $\{0\}$ at the bottom, and the subgroups of indexes 2 and 5 in the middle layer. Then, by the Fundamental Theorem of Galois Theory, the lattice of subfields between E and F is a diamond with subfields of indexes 2 and 5 in the middle layer.

19. By Example 7, the group is Z_6.

21. This follows directly from Exercise 21 in Chapter 25.

23. This follows directly from Exercise 31 in Chapter 24.

25. This follows directly from Exercise 40 in Chapter 10.

27. Since $K/N \triangleleft G/N$, for any $x \in G$ and $k \in K$, there is a $k' \in K$ such that $k'N = (xN)(kN)(xN)^{-1} = xNkNx^{-1}N = xkx^{-1}N$. So, $xkx^{-1} = k'n$ for some $n \in N$. And since $N \subseteq K$, we have $k'n \in K$.

CHAPTER 33
Cyclotomic Extensions

1. Since $\omega = \cos\frac{\pi}{3} + i\sin\frac{\pi}{3} = \cos\frac{2\pi}{6} + i\sin\frac{2\pi}{6}$, ω is a zero of $x^6 - 1 = \Phi_1(x)\Phi_2(x)\Phi_3(x)\Phi_6(x) = (x-1)(x+1))(x^2+x+1)(x^2-x+1)$, it follows that the minimal polynomial for ω over Q is $x^2 - x + 1$.

3. Over Z, $x^8 - 1 = (x-1)(x+1)(x^2+1)(x^4+1)$. Over Z_2, $x^2 + 1 = (x+1)^2$ and $x^4 + 1 = (x+1)^4$. So, over Z_2, $x^8 - 1 = (x+1)^8$. Over Z_3, $x^2 + 1$ is irreducible, but $x^4 + 1$ factors into irreducibles as $(x^2+x+2)(x^2-x-1)$. So, $x^8 - 1 = (x-1)(x+1)(x^2+1)(x^2+x+2)(x^2-x-1)$. Over Z_5, $x^2 + 1 = (x-2)(x+2)$, $x^4 + 1 = (x^2+2)(x^2-2)$, and these last two factors are irreducible. So, $x^8 - 1 = (x-1)(x+1)(x-2)(x+2)(x^2+2)(x^2-2)$.

5. Let ω be a primitive nth root of unity. We must prove $\omega\omega^2\cdots\omega^n = (-1)^{n+1}$. Observe that $\omega\omega^2\cdots\omega^n = \omega^{n(n+1)/2}$. When n is odd, $\omega^{n(n+1)/2} = (\omega^n)^{(n+1)/2} = 1^{(n+1)/2} = 1$. When n is even, $(\omega^{n/2})^{n+1} = (-1)^{n+1} = -1$.

7. If $[F:Q] = n$ and F has infinitely many roots of unity, then there is no finite bound on their multiplicative orders. Let ω be a primitive mth root of unity in F such that $\phi(m) > n$. Then $[Q(\omega):Q] = \phi(m)$. But $F \supseteq Q(\omega) \supseteq Q$ implies $[Q(\omega):Q] \leq n$.

9. Let $2^n + 1 = q$. Then $2 \in U(q)$ and $2^n = q - 1 = -1$ in $U(q)$ implies that $|2| = 2n$. So, by Lagrange's Theorem, $2n$ divides $|U(q)| = q - 1 = 2^n$.

11. Let ω be a primitive nth root of unity. Then $2n$th roots of unity are $\pm 1, \pm\omega, \ldots, \pm\omega^{n-1}$. These are distinct, since $-1 = (-\omega^i)^n$, whereas $1 = (\omega^i)^n$.

13. First observe that deg $\Phi_{2n}(x) = \phi(2n) = \phi(n)$ and deg $\Phi_n(-x) = \deg \Phi_n(x) = \phi(n)$. Thus, it suffices to show that every zero of $\Phi_n(-x)$ is a zero of $\Phi_{2n}(x)$. But ω is a zero of $\Phi_n(-x)$

means that $|-\omega| = n$, which in turn implies that $|\omega| = 2n$. (Here $|\omega|$ means the order of the group element ω.)

15. Let $G = \text{Gal}(Q(\omega)/Q)$ and H_1 be the subgroup of G of order 2 that fixes $\cos(\frac{2\pi}{n})$. Then, by induction, G/H_1 has a series of subgroups $H_1/H_1 \subset H_2/H_1 \subset \cdots \subset H_t/H_1 = G/H_1$, so that $|H_{i+1}/H_1 : H_i/H_1| = 2$. Now observe that $|H_{i+1}/H_1 : H_i/H_1| = |H_{i+1}/H_i|$.

17. Instead, we prove that $\Phi_n(x)\Phi_{pn}(x) = \Phi_n(x^p)$. Since both sides are monic and have degree $p\phi(n)$, it suffices to show that every zero of $\Phi_n(x)\Phi_{pn}(x)$ is a zero of $\Phi_n(x^p)$. If ω is a zero of $\Phi_n(x)$, then $|\omega| = n$. By Theorem 4.2, $|\omega^p| = n$ also. Thus ω is a zero of $\Phi_n(x^p)$. If ω is a zero of $\Phi_{np}(x)$, then $|\omega| = np$ and therefore $|\omega^p| = n$.

19. Let ω be a primitive 5th root of unity. Then the splitting field for $x^5 - 1$ over Q is $Q(\omega)$. By Theorem 33.4, $\text{Gal}(Q(\omega)/Q)$ $\approx U(5) \approx Z_4$. Since $\langle 2 \rangle$ is the unique subgroup strictly between $\{0\}$ and Z_4, we know by Theorem 32.1 that there is a unique subfield strictly between Q and E.

21. The three automorphisms that take $\omega \to \omega^4, \omega \to \omega^{-1}, \omega \to \omega^{-4}$ have order 2.

SUPPLEMENTARY EXERCISES FOR CHAPTERS 24-33

1. First observe that since
$$xy = (xy)^3(xy)^4 = (xy)^7 = (xy)^4(xy)^3 = yx, \quad x \text{ and } y \text{ commute.}$$
Also, since $y = (xy)^4 = (xy)^3xy = x(xy) = x^2y$ we know that
$x^2 = e$. Then $y = (xy)^4 = x^4y^4 = y^4$ and therefore, $y^3 = e$. This
shows that $|G| \le 6$. But Z_6 satisfies the defining relations with
$x = 3$ and $y = 2$. So, $G \approx Z_6$.

3. Let $|G| = 315 = 9 \cdot 5 \cdot 7$ and let H be a Sylow 3-subgroup and K a
Sylow 5-subgroup. If $H \triangleleft G$, then $HK = 45$. If H is not normal,
then by Sylow's Third Theorem the number of Sylow 3-subgroups
is 7 and therefore $|G/N(H)| = 7$ and $|N(H)| = 45$.

5. Observe that $K \subseteq N(H)$ implies that HK is a group of order
$245 = 45 \cdot 5$ (see Exercise 51 in Chapter 9 and Exercise 7 in the
Supplementary Exercises for Chapters 5-8). By Sylow's Third
Theorem (Theorem 24.5) and its corollary, K is normal in HK.
Thus, $H \subseteq N(K)$.

7. By the corollary to Sylow's Third Theorem K is the only Sylow
p-subgroup of H. But for any $g \in G$, we have
$gKg^{-1} \subseteq gHg^{-1} = H$. So, for all $g \in G$, $gKg^{-1} = K$.

9. Use the same proof as Exercise 51 in Chapter 9.

11. By Sylow's Third Theorem $n_7 = 8$ and by the Embedding Theorem
(Chapter 25) we know that G is isomorphic to a subgroup of A_8.
But A_8 does not have an element of order 21.

13. Let G be a non-Abelian group of order 105. By Theorem 24.6 we
know that Z_{15} and Z_{35} are the only groups of orders 15 are 35 and
by Theorem 9.3, $G/Z(G)$ is not cyclic. So $|Z(G)| \ne 3, 7, 15, 21$, or
35. This leaves only 1 or 5 for $|Z(G)|$. Let H, K, and L be Sylow
3-, Sylow 5-, and Sylow 7-subgroups of G, respectively. Now,
counting shows that $K \triangleleft G$ or $L \triangleleft G$. Thus, $|KL| = 35$ and KL is
a cyclic subgroup of G. But, KL has 24 elements of order 35 (since
$|U(Z_{35})| = 24$). Thus, a counting argument shows that $K \triangleleft G$ and

$L \lhd G$. Now, $|HK| = 15$ and HK is a cyclic subgroup of G. Thus, $HK \subseteq C(K)$ and $KL \subseteq C(K)$. This means that 105 divides $|C(K)|$. So $K \subseteq Z(G)$.

15.

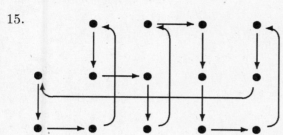

17. It suffices to show that x travels by a implies xab^{-1} travels by a (for we may successively replace x by xab^{-1}). If xab^{-1} traveled by b, then the vertex xa would appear twice in the circuit.

19.   **a.** $\{00, 11\}$

    b. $\{000, 111\}$

    c. $\{0000, 1100, 1010, 1001, 0101, 0110, 0011, 1111\}$

    d. $\{0000, 1100, 0011, 1111\}$

21. The mapping $T_v : F^n \to \{0, 1\}$ given by $T_v(u) = u \cdot v$ is an onto homomorphism. So $|F^n / \text{Ker } T_v| = 2$.

23. It follows from Exercise 18 that if C is an (n, k) linear code, then $C^{\perp}$ is an $(n, n - k)$ linear code. Thus, in this problem, $k = n - k$. Since $C = C^{\perp}$, to prove that $(11 \ldots 1)$ is a code word it is enough to prove that $(11 \cdots 1) \cdot u = 0$ for all $u \in C$. By Exercise 21, we know that $(11 \cdots 1) \cdot u = n/2 = 0 \mod 2$.